Qualitative Data Analysis with NVivo

Qualitative Data Analysis with NVivo

Pat Bazeley

First published 2007
Reprinted 2007, 2008

SAGE Publications Ltd
1 Oliver's Yard
55 City Road
London EC1Y 1SP

SAGE Publications Inc.
2455 Teller Road
Thousand Oaks, California 91320

SAGE Publications India Pvt Ltd
B 1/I 1 Mohan Cooperative Industrial Area
Mathura Road, New Delhi 110 044
India

SAGE Publications Asia-Pacific Pte Ltd
33 Pekin Street #02-01
Far East Square
Singapore 048763

Library of Congress control number 2006936346

British Library Cataloguing in Publication data

A catalogue record for this book is available from the British Library

ISBN 978-1-4129-2140-4
ISBN 978-1-4129-2141-1 (pbk)

Typeset by C&M Digitals (P) Ltd., Chennai, India
Printed on paper from sustainable resources
Printed in Great Britain by The Cromwell Press Ltd, Trowbridge, Wiltshire

Contents

Preface

Learning any new software program, even just updating to a radically changed version, is a daunting prospect. If you're like me, you groan at the thought, and wish for someone to stand by you to ease you through the process. This is what I attempt in this book – to be the guide beside you as you work though your first project in NVivo (using Version 7 or later). It's written for those who can't access personal training, and also for those who have, but who need then to think through how the huge confusion of ideas they brought home from their training session might be contextualized, applied and extended using their own data.

You will find within these pages a mix of rationale, instruction, methodology and examples, along with some friendly (experientially based) tips and warnings. While the topics covered in this book are organized approximately in a sequence you will find useful as you work through a project, use the contents page and the index to dip forward or back as needed. Analyzing qualitative data is never a neat, linear process!

Several people have patiently worked their way through each of the chapters as I have written them, each adding value in their unique way. I gratefully acknowledge the contributions of Barbara Bowers, Leonie Daws, Simone DeVore, Lynne Johnston, Lynn Kemp, Noriah Mohd Ishak, Lyn Richards, and Caroline Spencer, as well as the encouragement I have had from Patrick Brindle at Sage, London. I have learned a great deal also from the many students and clients I have worked with – projects from several of them feature in the examples I have given. And finally, my thanks go also to the team at QSR for creating software which is so flexible and so powerful, yet which maintains an intuitive and user-friendly interface. I hope my readers find it as useful for their research as I have found it for mine.

Chapter 1

Perspectives: Qualitative Computing and NVivo

This is a book for three kinds of learners:

- Those who prefer to learn by doing;
- Those who want to learn new tools for data management and analysis on a need-to-know basis;
- Explorers, who just want to play around and see what this software might do for them.

Through the course of undertaking a qualitative analysis project using this book, you will find out how to use one particular software program, NVivo.[1] On the way, you will find references, explanations and advice to help you understand what you are doing and why. And as you learn to 'drive' the software, you will also move along the road to completion of your project – a triple benefit!

In this chapter:

- Discover how use of software can support you in doing qualitative research;
- Read the story of how NVivo came to exist, and its intellectual history;
- Consider issues and objections people have raised about use of software for qualitative research;
- Get a sense of how NVivo will help you work with your data; and
- View an outline to guide your journey through the software and this book.

QUALITATIVE RESEARCH PURPOSES AND NVIVO

Researchers engage in projects involving interpretation of unstructured or semi-structured data for a variety of reasons. These might include exploration, description, comparison, pattern analysis, theory testing, theory building, or evaluation.

Methodologists routinely urge researchers to assess the fit between purpose and method (Maxwell, 2005; Richards & Morse, 2007), with the choice to use a qualitative approach being determined by the research question and purpose, rather than by prior preference of the researcher. Qualitative methods will be chosen in situations where a detailed understanding of a process or experience is wanted, where more information is needed to determine the exact nature of the issue being investigated, or where the only information available is in non-numeric (e.g., text or visual) form. Such investigations typically necessitate gathering intensive and/or extensive information from a purposively derived sample, and they involve interpretation of unstructured or semi-structured data.

How NVivo supports qualitative analysis

QSR International, the developers of NVivo, promise only to provide you with a set of tools that will assist you in undertaking an analysis of qualitative data. NVivo has been developed by researchers, with extensive researcher feedback, and is designed to support researchers in the varied ways they work with data. The use of a computer is not intended to supplant time-honoured ways of learning from data, but to increase the effectiveness and efficiency of such learning. The computer's capacity for recording, sorting, matching and linking can be harnessed by the researcher to assist in answering their research questions from the data, without losing access to the source data or contexts from which the data have come.

The average user of a software program typically accesses only a small proportion of its capabilities; this is no doubt true for users of NVivo also. Those using NVivo for a small descriptive project, for example, can work without having to learn complex procedures, while those undertaking complex analytical tasks can find the additional tools they need.

There are five principal ways in which NVivo supports analysis of qualitative data. Using software will assist you to:

- Manage data – to organize and keep track of the many messy records that go into making a qualitative project. These might include not just raw data files from interviews, questionnaires, focus groups or field observations, but also published research, other documentary sources, rough notes and ideas jotted into memos, information about data sources, and conceptual maps of what is going on in the data.

- Manage ideas – to organize and provide rapid access to conceptual and theoretical knowledge that has been generated in the course of the study, as well as the data which supports it, while at the same time retaining ready access to the context from which those data have come.
- Query data – to ask simple or complex questions of the data, and have the program retrieve from its database all information relevant to determining an answer to those questions. Results of queries are saved to allow further interrogation, and so querying or searching becomes part of an ongoing enquiry process.
- Graphically model – to show cases, ideas or concepts being built from the data, and the relationships between them, and to present those ideas and conclusions in visual displays using models and matrices.
- Report from the data – using contents of the qualitative database, including information about and in the original data sources, the ideas and knowledge developed from them, and the process by which these outcomes were reached.

There is a widely held perception that use of a computer helps to ensure rigour in the analysis process. Insofar as computer software will find and include in a query procedure, for example, every recorded use of a term or every coded instance of a concept, it ensures a more complete set of data for interpretation than might occur when working manually. There are procedures that can be used, too, to check for completeness, and use of a computer makes it possible to test for negative cases (where concepts are *not* related). Perhaps using a computer simply ensures that the user is working more methodically, more thoroughly, more attentively. In these senses, then, it can be claimed that the use of a computer for qualitative analysis can contribute to a more rigorous analysis. Even so, human factors are very much involved, and computer software cannot make-good work that is sloppy, nor compensate for limited interpretive capacity. As much as 'a poor workman cannot blame his tools', good tools cannot make up for poor workmanship.

Perhaps surprisingly, the tools described in this book are 'method free' insofar as the software does not prescribe a method, but rather it supports a wide range of methodological approaches. Different tools will be selected or emphasized and used in alternative ways for a variety of methodological purposes.

> We reiterate that no single software package can be made to perform qualitative data analysis in and of itself. The appropriate use of software depends on appreciation of the kind of data being analyzed and of the analytic purchase the researcher wants to obtain on those data. (Coffey and Atkinson, 1996: 166)

There are, nevertheless, some common principles regarding most effective use for many of the tools, regardless of methodological choices. For example, the labels used for coding categories will vary depending on the project and the methods chosen, but the principles employed in structuring those categories in a hierarchical coding system are common to virtually all methods where coding takes place. These common principles allow me to describe in general how you might use the various tools. It is then your task to decide how you might apply them to your project. Pointers to particular strategies which might suit particular methodological approaches are provided throughout this book, however.

If you are coming to NVivo without first meeting methodology or methods, then you are strongly advised to read first some general or discipline-based introductory texts. Then use the recommended reading lists in those, references in this book, or Google 'qualitative research' to further explore the methodological choices available to you.

THE NUD*IST-NVIVO STORY[2]

NUD*IST 1 was born in 1981 after Tom Richards set out, in 1979, to master programming in order to assist his sociologist wife, Lyn Richards, in managing the data files from a large neighbourhood research project. At the time, Tom was an academic, teaching logic at La Trobe University in Melbourne and just moving into computer science, while Lyn was a family sociologist, also teaching at La Trobe University. Lyn provides a rather delightful description of the problems she was experiencing with paper coding techniques, specifically when multiple copies of a segment of text had to be made and sorted into piles: a task that "was boring, time consuming, and not very rigorous, since dogs and babies were likely to mix with the precious paper segments" (Richards, 2005: 89). The particular experience of her two-year-old son crawling through the piles of data on the lounge room floor and eating a never-to-be-retrieved quote sparked the conversations which led to development of the program of which she referred to herself as 'mother', while noting (in conformation to contemporary mores) that it had a legitimate 'father'.

The software was designed with a dual database, most obviously evidenced in the first graphical interface versions. These showed two main windows on the screen: one was a window into a document system, which held all the 'raw' research data, and the other was a window into a coding system, which held the researcher's evolving knowledge about the data. Ideas and concepts drawn from the data were stored in 'nodes' which held references to the source text. This system of text referencing allowed the retrieval, from the documents, of all the text passages currently coded at the node (meeting a need to code-and-retrieve), but it did more. The separation of node from document was both innovative and pivotal; it is what has made possible manipulation and revision of categories while retaining links to the evidentiary texts (T. Richards, 2002). In doing so, it allowed for the emergent nature of knowledge gained through interpretive analysis. Additionally, and uniquely, nodes were organized in a hierarchical tree structure, a cataloguing system which enabled sorting (and thus classification) of the categories being derived from the data.

The needs of qualitative researchers to pursue leads in their data required, however, that a computer program be able not only to retrieve all the text on a particular topic, but also to find text related to a combination of topics through interrogative searches. Then, perhaps, that found text might also become data into a further enquiry relating it to something else – a revolving results-data-results process referred to as 'system closure' (T. Richards, 2002). Tom recognized, also, the value of being able to obtain the results of multiple comparisons in one query procedure – the ability to examine, for example, gender differences (male, female) across a range of attitudes (or experiences, or issues, or ...), or to identify which solutions were used in relations to which problems – hence the idea of matrix searches where the patterning of relationships between concepts represented by sets of nodes could be viewed in a 'qualitative cross-tabulation'. NUD*IST 1 (and later versions) supported 17 ways of interrogating coded data, allowing both logic based (Boolean) and fuzzy (proximity) queries, almost none of which had been possible using manual methods of coding and analysis.

Experience with issues raised by NUD*IST 1 led to the development in 1987 of NUD*IST 2, still on mainframe. Version 2.3, in 1990, took the software onto Macintosh, but the program still had a mainframe-style, scroll-mode interface, requiring for example that coding be done on paper and then transferred into the computer using typed instructions (in document 'X', code text units 20–23 at node (3 4)). Version 2 was the first made available to the public for purchase. (I bought version 2.3 early in 1991, with licence number 38!)

Development of a windowing interface resulted in NUD*IST 3, released on Macintosh in 1993 (with no mainframe version), and on PCs using Windows 3.1 soon after, in 1994. NUD*IST 3 allowed for on-screen selection and coding of document text, and was particularly characterized by the display of nodes as a visual tree. It also saw the addition of a series of processing refinements which allowed for editing of text units, the placing of restrictions on the scope of searches (which effectively reduced a step in repeated searching), and, some time later, the innovative ability to merge two projects into one. A command file system, for automated processing of routine coding and searching tasks, was available in earlier versions.

With the growing world-wide adoption of NUD*IST 3 it became necessary to move out from the corner of Tom's laboratory in the computer sciences building, and to establish a company – Qualitative Solutions and Research Pty Ltd – to handle program development and marketing. Qualitative Solutions and

Research became independent of La Trobe University in 1995, and was later renamed QSR International.

The release of NUD*IST 4, in April 1997, provided much greater flexibility in working with data stored at nodes. The concept of the 'free node' was born, a node placed outside the tree structure until (and if) an appropriate place in the hierarchy could be determined. More significantly, 'live' access to coded data, via the node, allowed for recoding of already coded material, without having to return to the original documents. Data, now recontextualized at nodes, could be further coded while viewing the node, and that coding would 'stick' to the text, regardless of from where it was viewed. This was a major advance in qualitative computing.

Also in N4, the ability to import demographic and other quantitative data directly from table-based software made for greatly improved efficiency in entering and using such data. Additionally, counts of documents coded at a series of nodes or in cells of a matrix, or of volume of text as represented by text units, could be exported from the program, facilitating its use for mixed methods research. Further refinements in versions 5 (N5-2000) and 6 (N6-2002) automated the formatting of text units, gave more flexibility in the handling and editing of text units, and made it easier to access and report on matrix results. N5 and N6 were actually released after NVivo as the programs of choice for large, repetitive, or highly structured projects (facilitated by command files).

The parallel release of NVivo 1 in 1999 met three specific needs of qualitative researchers (T. Richards, 2002): to apply character-based coding, to have the facility of rich (formatted) text available, and to be able to freely write or edit text, without invalidating earlier coding. Provision was made for linking to other media (of any sort), and to split the tasks being carried by nodes. A case nodes area was added alongside free and tree nodes; attributes with values replaced nodes for holding demographic and other quantified data; flexible sets of documents or of nodes replaced the use of coding to allow restrictions in (scoping of) searches. A visual modeller that allowed nodes (and other data items) to be viewed in any kind of relationship was added, to allow for concept mapping. Processes of coding and working with data became more visual and more flexible in NVivo, making it a program of choice for working in a fine grained way with data.

With NVivo 7, the two lines of software development were brought together in an entirely new database, to cater for researcher needs to undertake projects ranging from fine, deeply reflective analysis to analytic processing of larger volumes of text sources. In learning to use NVivo 7 and later versions, researchers still draw on the rich heritage of foundations laid in NUD*IST 1.

ISSUES RAISED BY USING SOFTWARE FOR QUALITATIVE DATA ANALYSIS

"Tools extend and qualitatively change human capabilities" (Gilbert, 2002: 222). Users of tools provided by NVivo may face opposition from those who express doubts about using software for analysis of qualitative data, or who simply have an aversion to technological solutions. Concern about how using a software program impacts on method is not limited to aging professors, and has attracted some debate at conferences and in the literature. If this is not an issue for you, feel free to move on to the next section of this chapter.

The development of software tools and advances in technology in general have had significant impacts on how research is done. These impacts are not limited to qualitative data analysis. The constantly expanding use of the web to provide access to data is now extending and changing the process of qualitative interviewing as well as the structure of surveys and survey samples. The widespread use of tape recorders in interpretive research has changed both level and kind of detail available in raw material for analysis, and as video recording becomes more common, data and method will change again. Tools range in purposes, power, breadth of functions, and the skill demanded of the user. The effectiveness with which you can use tools is partly a software design issue – software can influence your effectiveness by the number or complexity of steps required to complete a task, or by how information is presented to the user. It is also a user issue – the reliability, or trustworthiness, of results obtained depends on the skill of the user in both executing method and using software. The danger for novices using a sophisticated tool is that they can 'mess up' without realizing they have done so (Gilbert, 2002).

Historically, the use of qualitative data analysis software has facilitated some activities, such as coding, and limited others, such as seeing a document as a whole or scribbling memos alongside text. In so doing, early computer programs somewhat biased the way qualitative data analysis was done. Historically, also, qualitative researchers were inclined to brand all qualitative data analysis software with a capacity for supporting code-and-retrieve activity as being designed to support grounded theory methodology[3] – a methodology which has become rather ubiquitously (and inaccurately) associated with any data-up approach – with the implication that if you wanted to take any other kind of qualitative approach, software would not help.

Lyn Richards (2002) argued that the most radical *methodological* changes which came about with qualitative computing were not in what the computer could do, so much as the uses to which it could be put in furthering analysis. That coding could be done using a computer was not in itself a methodological advance, but the complexity and detail with which coding was made possible by computers, and the benefit of that in driving a complex and iterative data interrogation process, provided the basis for a radical shift in researchers' approaches to both coding and analysis.

Concerns about the impact of computerization on qualitative analysis have most commonly focused around four issues[4] which are discussed below:

- The concern that computers can distance researchers from their data;
- The dominance of code and retrieve methods to the exclusion of other analytic activities;
- The fear that use of a computer will mechanize analysis, making it more akin to quantitative or 'positivist' approaches; and
- The misperception that computers support only grounded theory methodology, or worse, create their own approach to analysis.

Closeness and distance

Early critiques of qualitative data analysis software suggested that users of software lost closeness to data through poor screen display, segmentation of text and loss of context, and thereby risked alienation from their data. The alternative argument is that the combination of tape recorders and software can give too much closeness, and so users become caught in 'the coding trap', bogged down in their data, and unable to see the larger picture (Gilbert, 2002; Richards, 1998).

Recent software has been designed on the assumption that researchers need both closeness and distance (Richards, 1998): closeness for familiarity and appreciation of subtle differences, but distance for abstraction and synthesis – and the ability to switch between the two. Closeness to data – at least as much as can be had using manual methods – is assisted by improved screen display, rapid access to data through documents or retrieval of coded text, identification of data in relation to source characteristics, and easy ability to view retrieved segments of text in their original context. Other tools are designed to provide distance, for example, tools for modelling ideas, for interrogating the database to generate and test theory, for summarizing results. These take the researcher beyond description to more broadly applicable understanding. Moving between these tools, from the general to the specific, and from the specific to the general, back and forth, exploiting both insider and outsider perspectives, is characteristic of qualitative methods and contributes to a sophisticated analysis.

Domination of code and retrieve as a method

The development of software for textual data management began when qualitative researchers discovered the potential for text storage and retrieval offered by computer technology. Hence, early programs became tools for data storage and retrieval rather than tools for data analysis, simply because that was what they were best able to do. The few programs that went beyond retrieval to facilitate asking questions about the association of categories in the data, particularly non-Boolean associations such as whether two concepts occurred within a specified level of proximity to each other (e.g. NUD*IST, Atlas-ti), were less rather than more common, and in these early stages were given special status as second-generation 'theory-building' programs (Tesch, 1990).

Computers removed much of the drudgery from coding (cutting, labelling and filing); they also removed the boundaries which limited paper-based marking and sorting of text.

> When recoding data involves laborious collation of cut-up slips and creation of new hanging folders, there is little temptation to play with ideas, and much inducement to organise a tight set of codes into which data are shoved without regard to nuance. When an obediently stupid machine cuts and pastes, it is easier to approach data with curiosity – asking "what if I cut it this way?", knowing that changes can be made quickly. (Marshall, 2002: 67)

Simply making coding more efficient was not, in itself, a conceptual advance from manual methods of data sorting (Coffey & Atkinson, 1996). Criticism that segments of text were removed from the whole, creating a loss of perspective, was frequently levelled at computer software (apparently without recognition that cutting up paper did the same thing, but with even greater risk of not having identified the source of the segment). Fears were expressed that computers would stifle creativity and reduce variety as code and retrieve became the dominant approach to working with data, to the neglect of extensive memoing, linking of ideas, holistic viewing of the text, and visualizing techniques.

Most problematically, the facility for coding has led to a kind of 'coding fetishism' – a tendency to code to the exclusion of other analytic and interpretive activities, which biases the way in which qualitative research is done (L. Richards, 2002). Historically, prior to the development of computer software for coding, much more emphasis was placed on reading and re-reading the text as a whole, on noting ideas that were generated as one was reading, on making links between passages of text, on reflecting on the text and recording those reflections in journals and memos, and on drawing connections seen in the data in 'doodles' and maps. Chapters 4 and 5 in this book provide suggestions for how to strike a balance when working with data using a computer, to combine coding with reading, reflecting, linking, noting, and doodling.

Viewed positively, ready retrieval of coded text allows the user to 'recontextualize' the data, to see it anew through the category rather than through the case or the document (L. Richards, 2002), and rapid access to original context of the segment has overcome the problem of disconnection. Furthermore, capacity to restructure a coding system facilitates the 'playing with ideas' referred to by Helen Marshall as an essential part of creative analysis, as does live access to the coded data so that, if desired, it can be reworked through further coding from the node.

Computers and mechanization

Fears that the computer, like Frankenstein's monster, might take over the analysis process and alienate researchers from their data stem in part from the historical association of computers with numeric processing of data, and in part from the computer's capacity to automate repetitive processes or to produce output without making obvious all the steps in the process. In addition, the association of computers with statistics, systematic processing, rigour and objectivity have all contributed to the (mis)perception of an association of computerization with positivism, whereas the majority of qualitative researchers are working within more interpretive paradigms.

The tendency of some authors to equate computer-based coding with mechanized coding procedures based on the search for keywords within the text adds to their and others' concerns about the computer alienating the researcher from the analysis process (*cf.* FQS, 2002). There are software programs designed to automate the coding process entirely, necessitating the development of complex dictionaries and semantic rule books to guide that process, but these are specifically designed for quantitative purposes, and the results of their coding are interpreted through the use of statistics without recourse to the original text (Bazeley, 2003). Keyword searches within *qualitative* analysis will almost always be secondary to interactive coding of the data, if they are used at all.

Automated coding processes have a place in handling routine tasks (such as identifying the speakers in a focus group), or perhaps in facilitating initial exploration of the data or for checking thoroughness of coding. These remove drudgery without in any way hindering the creativity or interpretive capacity of the researcher. Automated coding or keyword searches cannot substitute for interpretive coding, however; meaningful coding still needs to be done interactively (live on screen).

One of the goals of this book is to ensure that researchers using NVivo understand what the software is 'doing' as they manipulate their data, and the logic on which its functions are based – just as artisans need to understand their tools. Such metacognitive awareness ensures researchers remain in control of the processes they are engaging in and are getting the results they think they asked for (Gilbert, 2002), while the more aware and adventurous users can experiment with new ways of using NVivo's tools to work with their data – as the good artisan knows how to make his tools 'sing' to produce a creative piece of work.

Homogenization of qualitative approaches to analysis

You may hear researchers talk about 'doing qualitative' as if to imply there is just one general approach to the analysis of qualitative data, but 'qualitative' is not a method, and each attempt to define the characteristics of that general approach results in a different set of criteria. While there are some generally accepted emphases, there are also marked differences in qualitative approaches which stem from differences in foundational philosophies and understandings of the nature of social reality. The researcher must integrate their chosen perspective and conceptual framework into their choices regarding what and how to code, and what questions to ask of the data; software cannot do that.

Similarly, qualitative data analysis software has been talked about as if it supported just one qualitative methodology, or worse, that it created a new method. The range of disciplines and perspectives of people who will use this book will tell another story. In my own experience, I have taught NVivo to people in disciplines ranging from natural sciences through social sciences, business and the professions to creative arts. Each has been able to find tools in NVivo to support their work.

As you work through this book, you will find descriptions and examples illustrating how adoption of different methods and methodological approaches can influence your choices in using NVivo's tools. But for now, it's time to take a look at NVivo, to get a sense of how it might be useful for your research.

WHAT DOES AN NVIVO PROJECT LOOK LIKE?

Throughout this book I will be illustrating the principles and activities being discussed with examples from a number of my own projects, as well as some that have been undertaken by colleagues or students, and others from the literature. To give you an overview of the tools available for working in an NVivo project and of what you might be working toward, we'll stop to take a look at a project which is at quite an advanced stage. Because it is a moderately mature project, these instructions are not designed to show you how to make a start on working in your NVivo project, but rather, of what will become possible as you progress through your project.

As you read these instructions, and others in later chapters, you will encounter a number of special icons:

---

is the indicator that you are about to be provided with specific instructions for working in the software.

- ▸ indicates these are steps (Actions) for you to follow

✓ indicates a **Tip** or series of tips – handy hints to help you through.

! indicates a **Warning** – ignore at your peril!

indicates an illustration from the *Researchers* project.

The Researchers project

The *Researchers* project (based on real data and a serious research question) comprises interviews, focus groups and written texts from researchers at different stages of career development who have told of their developing involvement in research or their experience of being a researcher. The study explores what brings people into research, and what holds them there as committed, productive researchers. Originally developed and partially completed using NVivo 1 as a sample project for *The NVivo Qualitative Project Book* (Bazeley & Richards, 2000), it has now been reworked, added to, and set up as a progressively developed demonstration project for this book, using NVivo 7. You will find illustrative examples drawn from the *Researchers* project throughout the book, marked by ●. When you meet them, you might use those examples to check how the software works. For those who do not yet have their own data, the project provides data on which a technique can be practiced. The *Researchers* project is available for download from my web site (www.researchsupport.com.au) in several stages of analytic development.

For now, so we can view a project which is well underway, download the *Researchers tutorial* project file and save it in the My Documents area of your computer. If you are not able to access the *Researchers* project, then you might follow the same instructions using the *Volunteering* tutorial project that comes with the software.[5]

Setting up

If you don't already have the software on your computer, then your first step to using NVivo will be to install it either as a fully licensed or trial version on your computer.[6] Then you will be able to open the *Researchers tutorial*.

Use the *Getting Started* guide to find minimum computer requirements and detailed instructions for installing the software. These simply require that you insert the disk (or double click the downloaded software), have the license number that came with it ready to type into the appropriate part of the installation wizard, and follow the steps as they appear on screen. It is likely that you will be required, as part of this process, to install several supporting programs prior to installing NVivo itself: the installation wizard will guide you through the necessary steps. If you have a full license, you will then need to activate your license via the internet or by contacting QSR, in order to keep using NVivo beyond the 30 day trial period (check **Help > NVivo Help > Activating NVivo**, or see the *Getting Started* guide, if you're not sure what this means).

When you first open the software, you might find it useful to view the *Introducing NVivo* tutorial, accessed via the Help menu (NVivo tutorials). This provides a brief overview of the various elements in an NVivo project, using data from the tutorial provided with the software. Alternatively (or as well), the instructions below will introduce you to NVivo using the *Researchers tutorial* (or you could look at similar items in the *Volunteering* tutorial).

> If you have an earlier version (e.g., N6 or NVivo 2) on your computer, you do not need to remove that before installing NVivo 7 or later.

Introducing the NVivo workspace

Opening a project takes you into the project workspace from which you can access all the software's tools. Figure 1.1 illustrates the workspace and its components for the *Researchers* project. Apart from a menu bar and several toolbars, there are three areas, or views, for working in.

- From the ***Navigation View*** you can choose which component of the project you wish to work on. Here, you can organize items in your project into folders and sub-folders.
- The ***List View*** lists the contents of a selected folder. Most importantly, these can be opened from this view, and new contents can be added in this view. The contents of a folder can be sorted, and the ***properties*** (name and description) of items in the folder can be changed.
- The ***Detail View*** shows the actual content of an opened item, so that you can work with it.

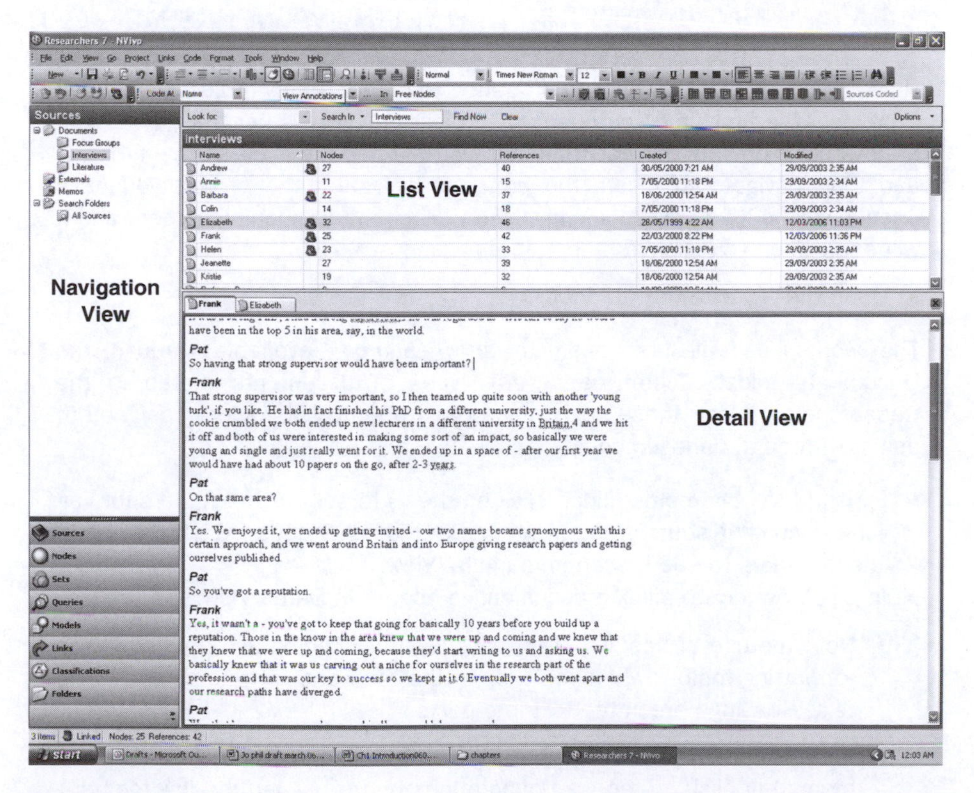

Figure 1.1 NVivo workspace showing menus, (adjusted) toolbars, and Navigation, List and Detail Views

The contents of menus, toolbars and views can change depending on which part of a project is active. In each view, context-sensitive menu items can be accessed from the right mouse button (**RMB**).

 Whenever you're not sure what to do or where to look for an action when you are working in NVivo, ensure your cursor or mouse pointer is pointing directly to the relevant item on your screen, and try the RMB as a first option for finding what you want.

As you explore the NVivo workspace using the *Researchers* project, you will gain some appreciation of how NVivo can assist with organizing and analyzing your data. Sources can be neatly filed; cases are identified with demographic and other details; ideas are recorded and appropriately linked to their sources; descriptive material and evidence for emerging understanding and ideas are captured in nodes; nodes are organized to facilitate querying the data so that research questions might be clarified, developed and answered; and for those who like to work visually, hunches, case analyses and emerging insights can all be explored in models.

A BRIEF EXCURSION THROUGH THE NVIVO WORKSPACE

NVivo opens to provide you with an option to create a new project or to open a project. The first time you open any project in NVivo, other than those created within it on your current computer, you will need to click on **Open Project** and then navigate to locate the project. Thereafter, the project you have opened will be listed as a **Recently Used Project** on the opening screen, and it will open when you click its title.

▶ Open the *Researchers tutorial*.

The workspace will first open to show ***Sources*** available through the Documents folder. There are several types of documents stored in the Researchers project: these have been organized into sub-folders designed to assist with data management.

▶ Expand the **Documents** folder (click on the +) to see further folders for various document sources (e.g., Interviews, Focus Groups). Click on one of these folders to see its contents in List View.

▶ In List View, double-click a document to open it in Detail View.

  ▶ Note the use of heading styles (the level of heading is identified in the formatting toolbar when you click in the text) and other markers in the text. Headings break the text into parts.

  ▶ If the text has blue highlighting, ask to **View > Annotations** (from the top menus) or click on the view annotations button in the View toolbar.

- Annotations are comments, reminders, or reflections on the text. Click on a blue highlight and the associated comment will be highlighted. Click on an annotation, and you will be shown the linked text.
- If words in the text have a red wavy line under them, it's not a spelling error, but an indicator for a ***see also link***. These links take you to other documents, files or passages which relate to the marked text. Choose **RMB** (right mouse button) > **Open To Item** to see the associated item.

- If the document has a linked memo (indicated by an icon next to the document in List View), open that by hovering (holding the mouse pointer) over the document name (or its text) and selecting **RMB > Links > Memo Link > Open Linked Memo**, OR, use **Ctrl+Shift+M** on your keyboard. Ideas and thoughts stimulated by the document are recorded in its linked memo.
- Note that you can have more than one document open at the same time: use the tabs at the top of the Detail View to change which document you are viewing. Close the document when you are finished with it (having too many things open at once impacts on available memory).

✓ Do not close a whole series of windows in rapid succession! (Take a breath between each one.)

Nodes provide the storage areas in NVivo for references to coded text. Each ***free*** or ***tree*** node serves as a container for what is known about, or evidence for, one particular concept or category.

- Click on the **Nodes** tab in the Navigation View to change the display from sources to nodes, and then on **Tree Nodes** to see the list of nodes used for coding in the later stages of this project. (Free nodes, which are not structured in any way, were used earlier, but these have since been moved into the more organized trees.)
- Expand one of the trees in List View by clicking on the **+** next to it, then select a node from within the tree by double-clicking. The text coded at that node will be displayed in the Detail View below (Figure 1.2). The source of each passage is identified, and the context from which a selected passage came can be accessed via the **RMB** (e.g., **Coding Context > Paragraph**; or **Open Referenced Source**).
- Select **Relationships** in the Nodes area of the Navigation View. You will then see in the List View, for example, that Elizabeth was encouraged by a number of different people in her journey to becoming a researcher. Additionally, a specific positive experience of encouragement can prompt interest in and perhaps a career in research. Not all relationship nodes have associated coding, but you can check the evidence gathered for those that do (indicated in the **References** column) by double-clicking on the relationship node.
- Select **Cases** to see the list of participants and all the data held for each in this project. Select a group participant, Ange or David, and double-click to see in the Detail View how each of their contributions in a focus group has been brought together.

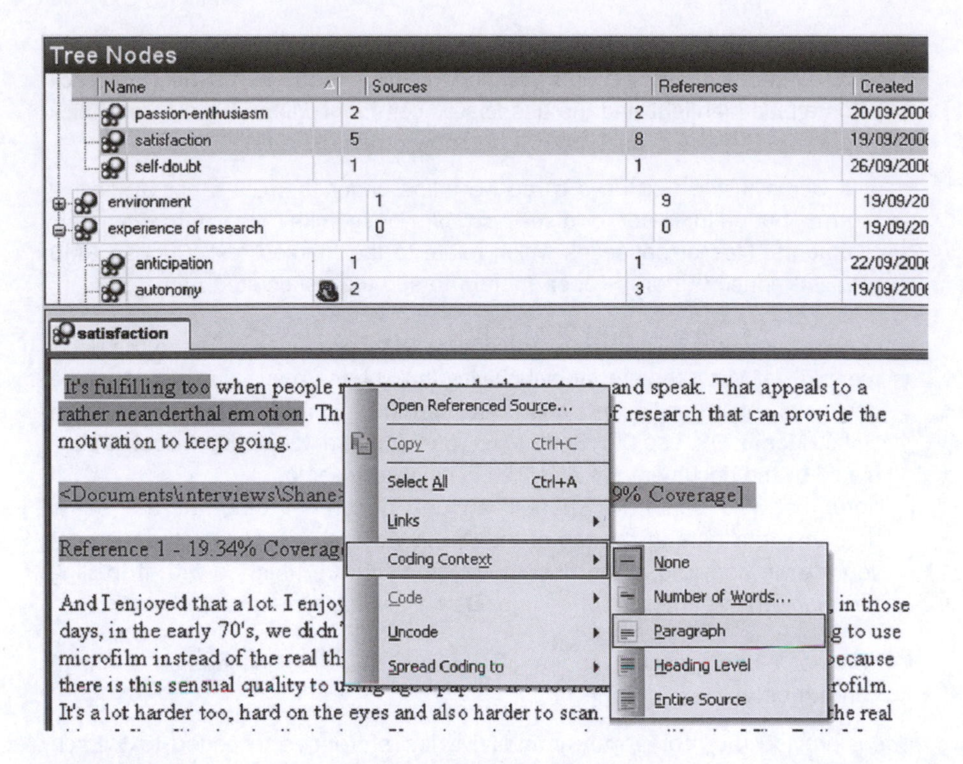

Figure 1.2 Tree Nodes with referenced text and context menu

- While in the Detail View for the case, from the View menu or using the toolbar icon choose **Coding Stripes > Nodes Most Coding Item** to see some of the nodes used in coding this case, and the lines of text coded by them. If you hover over the coding density bar, you will see a list of all nodes coding the adjacent text.

Cases can have ***attributes***, that is, a record of data known about the case that is recorded separately from the text generated by that case. This information will be used primarily to assist in comparing data for subgroups in your sample.

- View the attributes of a case by selecting a case, and from the **RMB** select **Case Properties** (or **Ctrl+Shift+P**). Click on the Attribute Values tab to see the values which have been assigned for this case.
- Attribute values for all cases can be viewed (and modified) in the Casebook (**Tools > Casebook > Open Casebook**). To record attribute data for Andrew, you first need to have made a case node for Andrew. To use the attribute data, you will need also to have coded Andrew's text data at his case node.

Sets in NVivo hold shortcuts to any nodes and/or any documents, as a way of holding those items together without actually combining (merging) them.

They are used primarily as a way of gathering items for use in handling queries, reports or models, or simply as a way of indicating that these items 'hang together' in some way (perhaps conceptually or theoretically).

▶ Click on the **Sets** tab in the Navigation View, click the + to see a list of sets, and then on a specific set to see its members.

Queries store questions you want to ask of your data. Queries might be about the occurrence of a word or words, about patterns of coding, comparison of groups, or some combination of these elements. They can be created and run once, or stored to use again with more data, or with a variation.
Results are stored along with queries. This is a temporary, read-only area for holding data found in answer to your queries.

▶ Click on the **Queries** tab in the Navigation View. Expand **Queries** to see its subfolders, and select the **Text searches** folder. Double-click the *exciting* query to see how a simple text search for the word exciting (and its derivatives) has been set up. Click **Run**. The results will open in Detail View. Use options from **RMB > Coding Context** to see the finds in context (e.g., paragraph). (If you press **Ctrl+A** first, you can see the context for all at the same time.)

▶ In **Results > List View**, double-click *encouragement from supervisor* to see how encouragement to engage in research might look when it comes from a dissertation supervisor. The text that is coded by both *encouragement* and *supervisor* will show in the Detail View. If you select the result node in the list view using your RMB, you can then select **Open Linked Query** to see how that result was obtained by requesting text coded at both *encouragement* AND *supervisor*.

▶ Results from comparative queries can be viewed also (e.g., *strategies* x *discipline*). These are useful for showing patterns of association between project items. Double-clicking on a cell will open the text for that cell.

Classifications is a work area for setting up attributes and their values, and types of relationships.

▶ Select **Classifications** in the Navigation View, to see how **Attributes** are set up (so values can be assigned to cases), and how **Relationship Types** are defined (for use in relationship nodes).

And finally (for now), ***Models*** are used in NVivo to explore, visually, ideas about how different project items might relate to each other.

▶ Select the **Models** tab in the Navigation View. There are a number of models which have been created in the *Researchers* project; these are identified in the List View. Select and view detail of a model by double-clicking its icon. To obtain a fuller view of the model, go to the Window menu and click on **Window > Docked** to undock the model window (allow a few seconds for it to open).

Save changes? While you have been looking at the *Researchers* project, you may have been warned that it was 15 minutes since the project was last saved, and asked if you wanted to save changes made to the project. This is NVivo's way of making sure that you are regularly saving changes to your project, in case of power failure or freezing. When you are working on your own project, it is strongly recommended that you save each time you are asked, unless you are simply experimenting, do not want to save your changes, or you are in the middle of an **Undo** operation.

Close the project by selecting **File > Close Project** or, if you want to quit working in NVivo for the time being, choose to **Exit**.

 If you would now like to convert your existing N4, N5, N6 or NVivo 1 or 2 project so you can explore it in the new NVivo, then search for the keyword **'legacy project'** in **Help** for information to assist you.

ABOUT THIS BOOK

The multiplicity of approaches to analysis of qualitative data poses particular problems for a book of this nature, with the purpose of walking you through a project: How best to organize and sequence the tasks and the introduction of different tools? As there is no standard pathway through a project, I don't expect everyone to work straight through this book.

The book is organized along broad sequential stages in working from raw data through conceptualization to strategies for analysis. You will need to step forward or back at various times to find the instruction, suggestions or discussion *you* most need for particular points in *your* project. Following the chapter-by-chapter sequence will take you through all the elements you need to consider: this may be the best general approach for someone new to the software. Those who already have some knowledge of the software from earlier versions may use the brief description of each chapter or the Table of Contents to identify where the major discussion of a topic they might be looking for is to be found within the chapters. A more detailed index is provided, of course, at the end of the book.

Instructions for various activities will be found throughout this book, but I will also provide the terminology needed to identify the appropriate topics in the Help files accompanying the software, so that these may be accessed for more detailed information. There are two kinds of Help provided within the software: *Using the Software* provides detailed instructions on how to perform various operations, while *Your Research in NVivo* suggests ways and provides examples for how you might apply particular tools. The Help files are both detailed and comprehensive, and they will be updated as the software is updated. They can be consulted, therefore, to resolve any discrepancies between my instructions or screen shots and the software that might result from ongoing updates to the software. Help can be accessed at any time, either via the **Help** menu or by pressing the **F1** key on your keyboard. Use the **Index**, or click on the **Search** tab, type in a keyword for what you are looking for and press **Enter** to find what you want. Clicking on blue hyperlinks will expand the notes to provide further detail and examples for many topics, and links at the bottom of the screens will take you to related topics.

Chapter outline

Chapter 2: Starting a project may surprise you as it suggests not that you start when you have data, but that you start a project in NVivo well before you have begun making data. Begin with a question and your thinking around that question, or perhaps with a theory, some literature and a model. By starting your project in NVivo earlier rather later, you will lay a sure foundation for working with data and verifying the conclusions you draw from them.

Chapter 3: Making data records provides guidance on making and managing data records in NVivo, including transcriptions, notes from literature and other documentary sources, external (non-file) sources, pictures and tables.

Chapter 4: Working with data outlines the day-to-day activities of reflecting on the text and coding as you work through your first documents. Suggestions are given for when and how to use memos, annotations and other links. Different ways of thinking about and doing coding are reviewed. Then, complete your initial analysis by creating a model of what you have learned from this case.

Chapter 5: Connecting ideas introduces you to two different ways in which you might think about making connections between the concepts you are working with: by sorting them into a hierarchical (tree-structured) classification system in which like types of concepts are stored together; or by identifying those which 'hang together' conceptually or theoretically using sets, relationship nodes, queries or models. Chapter 5 also offers a number of practical tips for managing the coding process.

Chapter 6: Managing data reviews how you can make use of folders, sets and cases in managing your data sources and to refine the questions you ask of your data. Discover how to store demographic and other kinds of information as attributes of cases, different ways of entering these into your project, and how to use them to make comparisons and examine patterns in your data.

Chapter 7: The 'pit stop' suggests you pause to view your data from the perspective of the category rather than your sources, and perhaps to refine the categories you have developed. You might also explore, code or investigate your data using text search, revisit the literature, and play with models using styles and groups, to build on your case knowledge and refine your theoretical thinking.

Chapter 8: Going further takes you beyond code and retrieve methods to explore applications of the analytic tools offered by NVivo, tools which will assist you in exploring cases, essences of experience, narrative and discourse, as well as in developing and testing theory.

Updates: Please check my website www.researchsupport.com.au for update information regarding instructions given in this book.

NOTES

1 Although this book is based on the use of NVivo (versions 7 and later), the tasks to be undertaken and the logic of use could well be applied to several other qualitative data analysis software programs (with some limitations and occasional opportunities). For a review of programs available, indicating strengths and weaknesses of the different programs on offer, and for access to developers' websites (most allowing downloads of sample or demo software) go to http://caqdas.soc.surrey.ac.uk

2 NUD*IST stands for Non-numerical Unstructured Data – Indexing Searching and Theorizing and so describes both the kind of data it was designed to work with, and the processes involved in working with that data. The full title of NUD*IST was replaced with just N for the release of versions 5 and 6 (also NVivo), in recognition of problems emanating from the dawn of an era of web searching.

3 Kelle (1997) traced the assumption that programs were written to support grounded theory to the need for a methodological underpinning for analysis, and grounded theory is one of the few methodologies where authors have been prepared to be explicit about what it is they actually do in analysis – although, as Kelle goes on to point out, "a closer look at the concepts and procedures of Grounded Theory makes clear that Glaser, Strauss and Corbin provide the researcher with a variety of useful heuristics, rules of thumb and a methodological terminology rather than with a set of precise methodological rules" (1997, #3.4).

4 One further issue of practical concern to researchers in money-strapped institutions is that they often will not support researchers' access to the software. The lobbyist might suggest to the relevant IT or Research Committees that statistical software be withdrawn from the quantitative researchers, or that administrators and libraries return to their card filing systems, for in refusing to provide access to qualitative data analysis software they are asking the ever increasing numbers of qualitative analysts to stay with pre-computer technology, when specialist software that will render their work more efficient and more effective is available.

5 Lyn Richards has written a series of interactive tutorials using the Volunteering project (the sample project supplied with the software), which some readers may wish to work through as an additional way of learning the software. They are available at www.sagepub.co.uk/richards. Additionally, Getting Started macromedia tutorials which illustrate features of and tools in the software can be accessed via the Help menu in the software (Help > NVivo Tutorials).

6 If you don't yet have the software, you can purchase it through the QSR web site (www.qsrinternational.com), or download a free trial version from the same site. You will then have 30 days to explore it before having to decide whether to buy or not.

Chapter 2

Starting a project

> There are many ways of putting off an "official" start [to a project]. Rethink the topic, rewrite the proposal, redo the preliminary literature review. All of these are proper processes in qualitative research, but they are wasted if they are not treated as part of the research. Explorations of the topic, the proposal, the literature, are data. The project is underway once these are happening. From the beginning, then, we need ways of storing and exploring, reworking and revising those early thoughts and rethinks, tentative ideas, insights, despairing memos, or discoveries in unrelated literature. From this oddly assorted treasure trove of early materials, we build a project up, as a bower bird builds a nest. (Bazeley & Richards, 2000: 10)

Starting is difficult, but also very satisfying. The suggestions in this chapter will not only get you started on ways of exploring and storing early thoughts and on using the software to do so, but when completed will give you a sense that you have really achieved something and that you have moved forward in your thinking about your project.

Varying approaches to analysis among different qualitative traditions impact on task choices very early in a project, the sharpest differences being between theory-generating and theory-testing approaches. While different approaches demand different starting points, there are some common denominators in the kinds of tools that might be used at this stage, and additional tools for those who have, *a priori*, a clear idea of what they are seeking to find.

In this chapter:

- Learn how you can start using software at an early stage in your project, well before 'analysis of data';

- Create a journal for the project as a whole, in which you might begin by noting your starting questions and assumptions for your new project;
- (optionally) Set up a starter coding system;
- Create a model of what you think you already know/assume/expect to find with regard to your topic;
- Learn how you can save your work securely.

STARTING

– with a question

Qualitative research often begins with a vaguely defined question or goal. It may well begin "with a bit of interesting 'data'" (Seale *et al.*, 2004: 9). Something in one's social or working environment excites interest, and investigation is begun. Investigation may begin in the library, or by observing 'the field', with some exploratory discussions with relevant people, perhaps with reflection on personal experience. Visualization techniques (concept maps) and thought experiments can help to clarify what might be useful questions (Maxwell, 2005). All these initial explorations serve to refine the question, so that more deliberate ('purposive') data gathering can occur. All become part of your data, and all can be managed within NVivo.

Record these starting questions as you set out. They will help you to maintain focus as you work, and later to evaluate the direction you are moving in. Keep notes also about any random (or less random) thoughts you have around those questions as you read, discuss, observe, or simply reflect on the issues they raise, and date these. Keeping a record will allow you to keep track of your ideas and to trace the path those ideas have taken from initial, hesitant conceptualization to final, confident realization.

– with assumptions

Record assumptions you are bringing to the project, too. Maxwell (2005) recommends creating a 'researcher identity memo' to explore not only personal goals, but also to recognize assumptions and experiential knowledge, as a way of developing the kind of "critical subjectivity" in which "we do not suppress our primary experience; nor do we allow ourselves to be swept away and overwhelmed by it" (Reason, 1988: 12).

The belief that an inductive approach to research requires that researchers come to their data without having been influenced by prior reading of the literature in their field and without bringing any theoretical concepts to the research is generally no longer seen as realistic nor broadly supported, although Charmaz (2006) advocates deferring *writing* the literature review until after analysis and theory development. Rather, Strauss and Corbin (1998: 47) declare that:

> Insights do not happen haphazardly; rather they happen to prepared minds during interplay with the data. Whether we want to admit it or not, we cannot completely divorce ourselves from what we know. The theories that we carry in our heads inform our research in multiple ways, even if we use them quite unselfconsciously.

while Kelle (1997: #4.2) argues that previous knowledge is a crucial prerequisite of gaining understanding:

> Qualitative researchers who investigate a different form of social life always bring with them their own lenses and conceptual networks. They cannot drop them, for in this case they would not be able to perceive, observe and describe meaningful events any longer – confronted with chaotic, meaningless and fragmented phenomena they would have to give up their scientific endeavour.

In a lighter vein, Lyn Richards has been heard to suggest that those who declare they have no prior assumptions will "walk them in on their boots".

Thus, rather than deny their existence, recognize them, record them, and become aware of how they might be influencing the way you are thinking about your data – only then can you effectively control (or at least, assess) that impact.

– with data

Not all qualitative projects are of the data-up, emergent-theory type. Nor are all based primarily on data derived from transcribed interviews or focus groups. Consider the possibilities of:

- Starting with observations of self and others – in which case, field notes or diary records will play a significant early role. Adapt the instructions for creating a project journal (described below) to create documents in which to record your observations and self-reflections. The next chapter on making data records will give further guidelines for setting out and managing field notes.
- Starting with a review of literature or other data already in the public sphere such as newspapers, novels, radio, web, archived data (Silverman, 2000). These can provide valuable learning experience as you master the software and analysis strategies. Again, the next chapter on data records has some suggestions for incorporating articles, notes from literature, or various types of existing documents into your project.
- Starting with a theory, which means you already know what kinds of things you will be looking for in your data (whilst keeping yourself open to new ideas, of course). This means you are likely to want to create a starter coding system – see below for guidance on how to do this.
- Starting with open-ended survey responses. Depending on the extent, significance and role of this data, keeping a detailed journal may not be so critical here, and your model may be more statistical than qualitative.

Document layout for these responses is critical, however, especially if there are multiple questions or multiple cases within any one document, so take particular notice of formatting issues covered in the next chapter.

STARTING WITH SOFTWARE

Your project begins from the time you start asking questions – from the perhaps casual thought that 'X' might be something interesting to investigate. This is also a good time to start using software!

- Early use of software ensures that you don't lose precious early thoughts. Indeed, writing even rough notes helps to clarify thinking as you plan your project.
- Starting early, if you are still learning software, will give you a 'gentle' introduction to it and a chance to gradually develop your skills as your project builds up. This is better than desperately trying to cope with learning technical skills in a rush as you become overwhelmed with data and the deadline for completion is looming.
- Starting with software early acts as a reminder that data collection and data analysis are not separate processes in qualitative approaches to research. So start now!

Two of NVivo's tools are useful to most researchers at this beginning point. A new blank document created within NVivo will serve to become a project journal, an ongoing record of questions, ideas, and reflections on the project. And construction of a conceptual model will show what you already know, point to what you yet need to know, and perhaps assist in identifying steps on the pathway to finding out. Additionally, for those who are testing established theory, or who have a clear idea of where they are going in their project, it may be useful also to create a starter coding system.

Setting up NVivo

But first, the software needs to be installed, if it isn't already (*cf.* Chapter 1). You might also wish to modify some program preferences to suit the way you like to work.[1]

SETTING PREFERENCES

- ▸ It is a good idea to check application defaults for heading styles[2] (and other features, including language options for non-English speaking users) and set them to the way you want them *before* you create a project. Otherwise, if you have already begun a project, you will have to set up the styles again for that project as well, under File > Project Properties. You can access **Options** from the **Tools** menu as soon as you launch the software (Figure 2.1). If you are unsure how to make changes, guidance on setting options is provided under the topic **Setting Application Options** in **Help**.

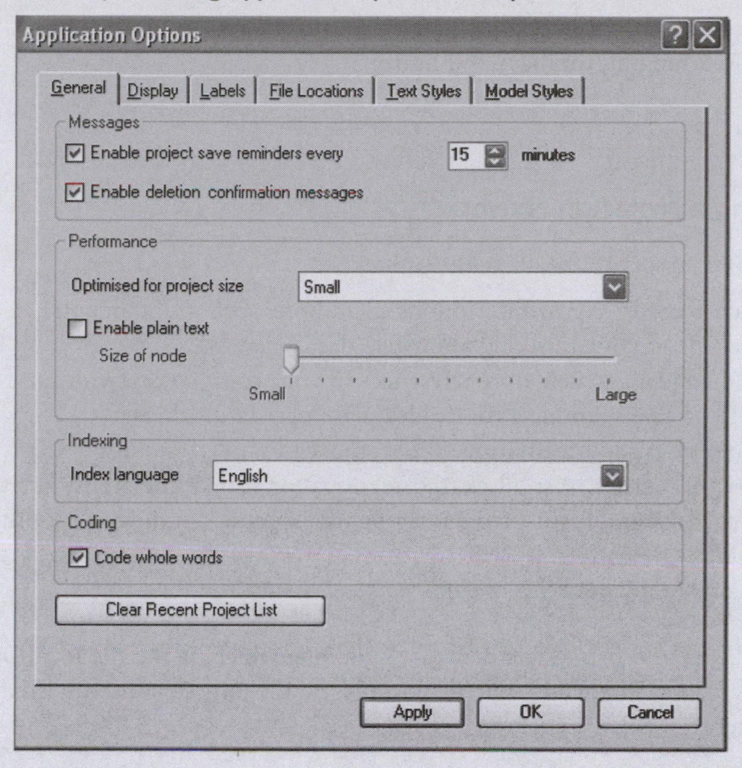

Figure 2.1 Setting application options

- You might also want to rearrange the toolbars so that they take up less of your screen space (Figure 2.2). For example, on my desktop (screen resolution 1280 × 1024) I removed just a few buttons I don't use (click on Toolbar Options at the end of any toolbar) and then moved the toolbars (drag on at the beginning of any toolbar) so that they occupied two lines rather than four. For the lower screen resolution on my laptop (1024 × 768) I removed a range of buttons that I'm unlikely to use while I'm working in NVivo from the Main and Edit toolbars, and closed the Grid toolbar. Toolbar positions are remembered next time you open the software.

✓ To change your screen resolution so that you have a larger or smaller work area: right-click on your desktop, select Properties, choose the Settings tab, and click on More or Less in the Screen Resolution box. Note that having a larger resolution means you may need stronger glasses! You need at least 1024 × 768 when using NVivo.

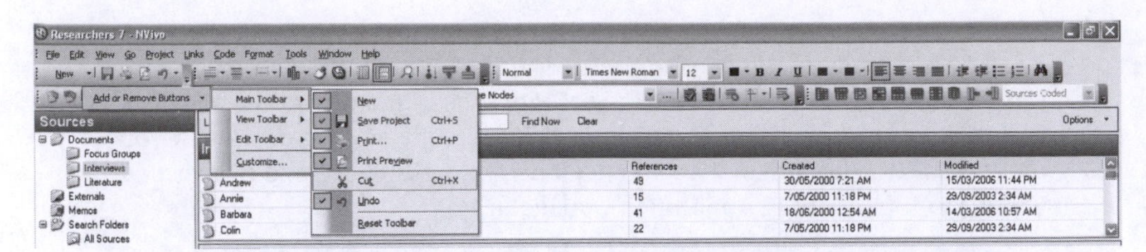

Figure 2.2 Rearranging toolbars and buttons

Creating a project in NVivo

An NVivo project typically comprises:

- data records (e.g. transcriptions, field notes, other documents),
- records of your thinking about the data (memos),
- coding items which store references to your data (so that you can retrieve all you know about a topic, idea, case or relationship),
- variable-type information about the cases in your study (e.g. demographic details, responses to categorical or scaled questions, dates),
- records of and results from interrogative queries conducted on your data, and
- models showing relationships between items in your project.

All of this is held in a single database-style file, which, if the file location options have not been changed, will be located in the **My Documents** area of your computer.

CREATING, NAMING, AND STORING PROJECTS

Creating a new project

- ▶ Creating a project in NVivo is as simple as clicking on **New Project** at the base of the Welcome screen, and typing a **Title** for the project into the New Project dialogue. Add a Description if you wish, to help identify this particular project.

- ▶ If you need to set a password and/or access rights to the project, this is done once the project has been created, by accessing **File > Project Properties**. Unless you have a compelling reason to do so (or a faultless memory), it is generally safer to *not* set a password for the project.

When you have created a project, it will be added to the **My Recent Projects** list on the Welcome screen, and in future you will be able to reopen it with a single click on its name.

Saving the project

As soon as a new project is created, it is saved.

- You will be asked every 15 minutes whether you wish to save changes to your project. Unless you were about to undo an action, click **Yes** each time, to ensure your work is not lost.
- This time lapse can be changed via **Tools > Options**. The pop-up reminder does briefly interrupt what you are doing, so more frequent may not be better. Less frequent, of course, carries obvious attendant risks regarding loss of work should the power go off or the program close for some reason.

Renaming a project

- A project can be renamed by going to **File > Project Properties**. To avoid confusion, you should also change the filename to match the project name (*after* you have closed the project!). The project name is a Windows registry entry recognized by the software: this is what shows in the Recent Projects List and at the top of the NVivo workspace. The filename is what you will see in My Documents and in Windows file navigation dialogues.

✓ To open a project after renaming or moving it, you will have to choose **Open Project** and then navigate to find it, rather than click on its name.

Deleting a project

Just in case you want to start over!

- Projects are deleted through the regular Windows file system (Windows Explorer, My Documents, My Computer). Project names will, however, persist in the Recent Projects list on the NVivo Welcome screen even after project files have been deleted in Windows. Next time you choose a deleted project name you will be asked if you want to remove it from the list, or, clear the entire list through **Tools > Options > Clear Recent Project List.**

Clearing the list does not, in itself, delete projects. Click on **Open Project** to find your current projects in My Documents (or wherever you saved them). The *Volunteering* project will be located in Shared Documents.

One or many projects?

Your research project may have a number of components with data generated from different sources (rural/urban; Companies A, B and C) or at different phases of the project (pilot phase/main data collection; wave 1, 2 and 3 of interviews), or with data of different types (e.g., notes from literature, observations, interview transcripts). NVivo provides data management tools with which you can either compare or isolate different components of your project (*cf.* Chapter 6). What this means in practice is that it is best to incorporate all those components into a single NVivo project, rather than making separate projects for each component. Having everything together in one NVivo project will allow you to gather together everything you know on any topic, regardless of source, and to make instant comparisons across different sources, phases, types of data or cases. If you wish, you will still be able to interrogate just one component of the data by placing relevant sources within a specific ***folder*** of documents or cases, or by identifying that component as belonging to a defined ***set***.

The one possible exception I sometimes make to the one-project rule is for topics on which I am likely to gather a *large* amount of reference material (articles, or notes from literature generally) as well as fresh interview or other data. In such cases I would consider putting just the reference material into a separate project. If it is literature which I am likely to use for other, related projects, or want to keep coming back to again and again without necessarily wanting to access other data, then I would certainly put it in a separate project.

A coding system developed while reviewing literature can be imported (without text references) into a new project in which you will be considering fresh data. Check the **Import Project** options under the File menu.

Will my project get too big for one file?

Unlikely – and if it does, it might be time to question whether this is a project you should be tackling qualitatively, especially if you hope to remain sane! Traditionally, most qualitative methodologists advise working with quite small sample sizes, in the order, say, of 10–50 interviews, or 5–10 group discussions, although ethnographic projects which run over years tend to accumulate significant volumes of field notes and interview data. Increasingly there is an expectation to handle a larger amount of data, however, perhaps because of the need to stratify a sample, because the project is multisite and the work of different teams has been merged, or because it is a longitudinal study. NVivo 7 has been designed to handle a virtually unlimited volume of text, although of course, it works more efficiently with a less than unlimited amount (check the FAQ section for NVivo 7 on the QSR website for potentially useful information on this issue).

If you do have 20,000 newspaper articles or 12,000 responses to an open ended question from a survey, for example, it is more appropriate to take a sample of these for detailed analysis than to try to code all in NVivo. You might then generate from that analysis a list of categories of responses (with an understanding from the qualitative analysis of what those are likely to mean) to use as a basis for category coding the entire sample directly into a spreadsheet or statistics program. Category coding will allow you to create counts, cross tabulations (pivot tables) or undertake other statistical analyses with numeric variables gathered at the same time.

Similarly, if you have, say, a *very* large amount of documentary data, then perhaps what is needed to give a first cut on the data is simple word counting or quantitative content analysis software; software which will provide key words in context (KWIC);[3] or use Excel to summarize and sort key points under predetermined headings for each document (Bazeley, 2006). NVivo is versatile, but it is not the answer to every data-processing need!

If you are working with a large amount of data in NVivo, change the default option on project size through **Tools > Options**. Help suggests that having over 500 sources constitutes a large project, but this will depend on your computer capacity as well.

Creating a journal

Qualitative researchers typically keep a journal to document how they have moved from initial forays in their project to arrival at their conclusions; hence some refer to the journal as an audit trail for the project. Lyn Richards (2005), in *Handling Qualitative Data*, compares the journaling process to keeping a ship's log with its careful account of a journey, and provides detailed suggestions about what might be recorded there (see especially pp. 23–4 where she outlines four questions to answer whenever the research changes direction). Without such a record, it will be difficult to keep track of when and how insights were gained and ideas developed, and it may be difficult to pull together the evidence you need to support your conclusions. Without it, too, precious, fleeting ideas will become forgotten as the data marches in, the next task is upon you, or the complexity of it all begins to overwhelm. Unlike the ship's log, however, the journal can be a private document: you might also record your frustrations and your joys as you work through your project. Perhaps the best advice of all, as you focus on ideas and your responses to them (rather than dry description) is to enjoy the journaling task – write freely without worrying about formality of style or 'correctness' of thoughts. Writing "often provides sharp, sunlit moments of clarity or insight – little conceptual epiphanies" (Miles & Huberman, 1994: 74).

Creating a journal requires that you have a document to record into. If your document is within NVivo, it will always be available as you are working in the project for you to add to, and you will be able to establish ***see also links*** from your thoughts to specific data or other evidence which prompted or supports those thoughts (*cf.* Chapter 5). No more coloured tags hanging off the sides of pages to help you find those insightful ideas! As for any document in the project, you will be able to code the journal as you write it, making it easy to retrieve the ideas you have had on any topic. Use NVivo's date and time stamp on journal entries, to help with the auditing process.

CREATING A JOURNAL

Creating a new, blank document

Documents in NVivo can be imported (*cf.* Chapter 3), or they can be created within the program. For now, we're going to create a working document in the program so it is there as a kind of 'scratch pad' for ideas and thoughts. Because it is a journal rather than a data source, we will be creating it as a memo.

- Select **Memos** from within the Sources area in the Navigation View, then right-click in the List View. Choose **New Memo**.
- **Name** the memo, provide a **Description** if you wish, and click **OK**. The new document will appear in the List View. Double-click to open it in Detail View, for adding text.

✓ If you place an underscore at the beginning of a document name, for example, **_Journal**, then it will always appear at the top of any alphabetically sorted list. This is especially useful for something like a project journal, making it faster to access.

Writing in the document

Working with your journal in the Detail View, you can now begin recording the questions, assumptions or other ideas you are bringing to the project. The following prompts might help here:

Why are you doing *this* project?
What do you think it's about?
What are the questions you're asking, and where did they come from?
What do you expect to find and why?
What have you observed so far?

Make use of the editing tools available through the formatting toolbar as you work.

- Use **Ctrl+Shift+T** to insert the date and time, or locate it under **Format > Insert > Date/Time** (time stamps do not automatically become codes).

- Select fonts and use colour to add emphasis (you cannot automatically code text on the basis of colour). For example, I use red font for identifying questions to think about.
- Use heading styles to identify parts of the text. Note that simply making something bold or larger does not identify it as a heading: it is necessary to select a style from the Styles slot in the formatting toolbar.

✓ Because heading styles in NVivo do not revert to Normal style when you take a new line, it is easier to click and add the style *after* you've written the next paragraph.

✓ Be consistent about using the same level of heading for the same kinds of things.

Saving your journal

Documents are saved along with the project as a whole in NVivo, that is, you do not save a document as a separate entity, even if you are closing the document. If, however, you are anxious to ensure that what you have just written is not lost, then choose **File > Save Project** (more on saving and backing up, below).

More than one journal?

Some researchers recommend setting up separate journals for different purposes in a project (for example: Miles & Huberman, 1994; Schatzman & Strauss, 1973). Separate journal documents could be used to record day-to-day thinking or tasks, for theoretical issues, and for methodological issues. I'm not so confident I can separate out my rather scattered thoughts so neatly into separate journals, and tend to start by keeping just a single document running. If you glance at later examples of the journal for the Researchers project you will find a broad mix of content, ranging from records of theoretically oriented conversations with colleagues, through reflections on concepts and experimentation with ideas, to simple 'to do' lists. Most content is dated (it should be), and is tagged with headings and codes to assist later review. In projects where I have created an additional journal, this has been either to record reflections on methods, or to record a brief summary of the main issue raised by each case (especially in projects with a lot of small documents supplementing quantitative data).

You might also need to refer to your research proposal. You can import this as a memo document, or you can create a hyperlink to it from your project journal, so that you can access it easily when you need to (work through the Links options on the RMB, or *cf.* Chapter 4).

Creating a starter coding system

The topic of coding is discussed at length in Chapter 4, *Working with data*. Work though this section only if: (a) you already have some understanding of how nodes work, and (b) you can identify concepts or categories from your questions or your conceptual or theoretical background which you know will be important in your data. In such cases, it can be useful to have a starter list of ***nodes*** – as long as you're prepared to change the nodes should you find they don't fit your data. *There is, however, absolutely no presumption that you need to make nodes at this point, before you begin working with your data.*

Use of nodes will be discussed in greater detail, along with coding, in Chapter 4. For the moment it suffices for you to be aware that nodes provide the storage areas in NVivo for accessing coded text. NVivo provides several types of nodes for keeping track of ideas and for organizing data. For most purposes, you are most likely to start with ***free nodes***, which do not assume relationships with any other concepts. Free nodes allow you to capture ideas without imposing any structure on those ideas, so they are particularly useful to use at the beginning of a project. Alternatively, it is possible to set up a hierarchical (tree) structure ahead of time, but only if you already know what 'kinds of things' you will be dealing with (e.g., a range of people, a set of attitudes, a selection of events, or actions, or places or times) *and* you understand the principles by which tree nodes should be organized – something that usually comes about only through experience in working with the program and with your data (*cf.* Chapter 5).[4]

The methods suggested here for making nodes are designed to be fast, suitable for when you are not primarily concerned with coding data at the same time. In contrast, for most of your work with data, creating nodes will be associated with concurrent coding.

OPTIONS FOR CREATING PRELIMINARY NODES

Create free nodes without coding

- From the Main toolbar (top left), choose **New > Free Node**

  OR,

  In the **Nodes** Navigation View, choose **Free Nodes**, then right-click in the List View area below existing nodes and select **RMB > Create Free Node.**

- In the Node Properties dialogue, provide a name (add a description if wanted).

✓ In the (Free Nodes) List View, use the keyboard shortcut **Ctrl+Shift+A**, then **Tab** and **Enter** on your keyboard, to avoid using the mouse altogether for making a list of nodes.

Create 'in vivo' codes from journal text

Where you have recorded questions or outlined existing ideas with concepts embedded in the terms used, use the 'in vivo' coding tool to capture those concepts and turn them into nodes. This will ensure that you have nodes designed to hold the data you need.

- From **Sources**, open your journal in Detail View, then change to **Nodes** in the Navigation Pane.
- Highlight the word or phrase in the text you wish to use as a node title, and click on the 'In Vivo' button in the coding toolbar. The node you have created will appear listed as a Free Node.

A node created in this way will also code just the word or phrase highlighted to make the node. It will not capture surrounding text, and it will *not* automatically find that word in other text. If, however, you delete the coded text from the document (or the whole document), the node will remain.

✓ Unless you have changed the options, NVivo will code whole rather than parts of words.

✓ If you already have an on-line list of potential node titles, import it and make free nodes of them by using the in vivo button on each one. Then delete the document from the project. The nodes will remain, ready for coding.

Creating nodes with coding

These methods are most appropriate when you want to code text at the same time as creating a node:

- Highlight the text to be coded.
- Click in the coding slot on the Coding toolbar, type in a name, and press **Enter** (or click **Code**). The node will be created in the designated area (Free) and the text will be coded (Figure 2.3);

  OR, **RMB > Code > Code Selection at New Node**;

  OR, press keyboard shortcut: **Ctrl + F3**.

! If you use the RMB option to code, watch that you don't accidentally choose to code the whole source.

✓ It is inconvenient at this stage to have some nodes in the Free area, and others in the Tree area. It is recommended, therefore, that you create them all in the Free area for now.

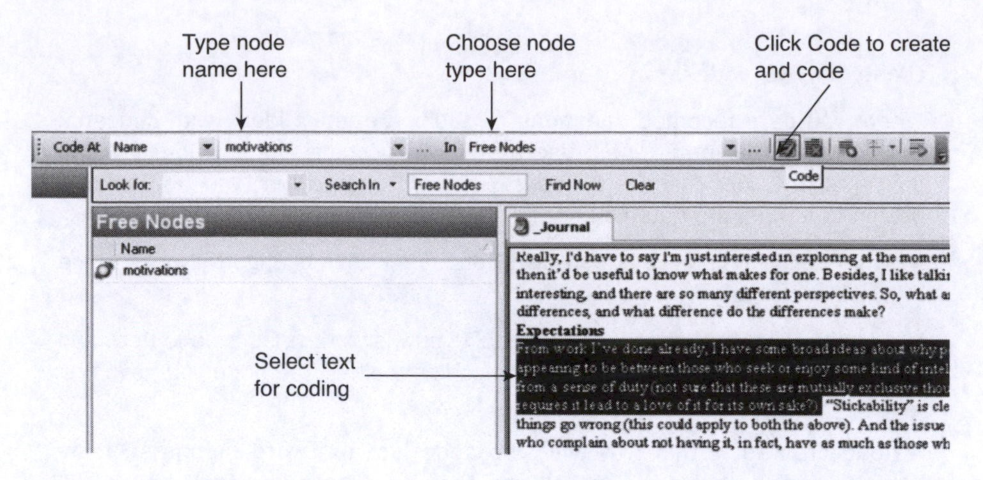

Figure 2.3 Creating nodes using the coding toolbar

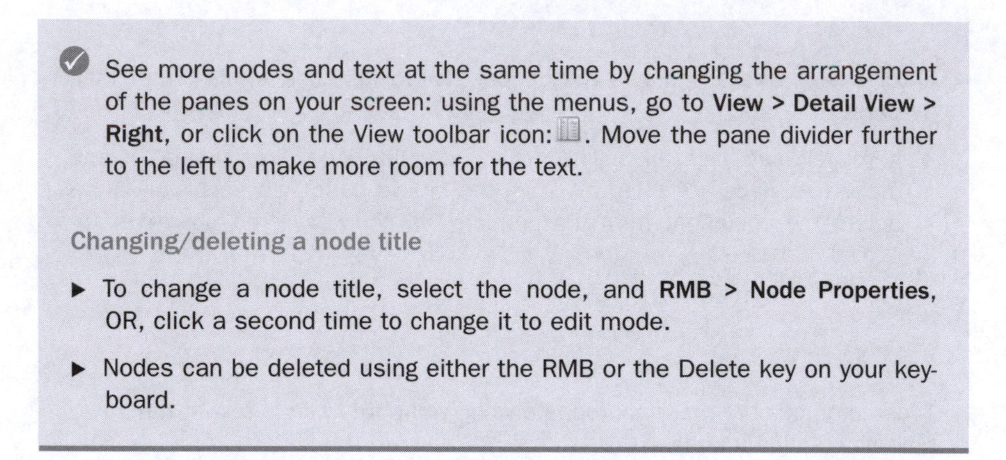

See more nodes and text at the same time by changing the arrangement of the panes on your screen: using the menus, go to **View > Detail View > Right**, or click on the View toolbar icon: . Move the pane divider further to the left to make more room for the text.

Changing/deleting a node title

- To change a node title, select the node, and **RMB > Node Properties**, OR, click a second time to change it to edit mode.
- Nodes can be deleted using either the RMB or the Delete key on your keyboard.

Creating a model

Sketching your ideas about your project at this stage is a particular way of journaling what you think it is you are asking or doing – great for those who prefer to think and work visually and, I've discovered, even for those (like me) who struggle to work visually. Miles and Huberman (1994: 91) confidently assert that "You know what you display". Maxwell (2005) argues strongly for creating an early concept map to help clarify the conceptual framework or theoretical underpinning of a study. In NVivo, conceptual maps, flow charts, or purely exploratory diagrams can be created using the modeling tool, and are generically referred to as ***models***. It doesn't matter whether or not you have nodes yet; put your concepts in a model anyway, and note observed associations or explore possible theoretical links.

Models serve multiple purposes during a qualitative research project. Just now modeling will provide a record of where you started from and what assumptions you have brought to the project. It may also assist with clarifying your research questions and planning your data collection.

So for now, use the NVivo modeler to make a diagram of the concepts, relationships or patterns you *expect* to find. If you find it a struggle to develop a concept map, then try some of Maxwell's (2005: 52) four suggestions, based on advice from Strauss (1987) and Miles and Huberman (1994):

- think about the key words you use in talking about the topic, or in things you've already written about your research;
- take something you've already written and map the implicit theory within it;
- take one key concept or term, and think of all the things that might be associated with it;
- ask someone to interview you about the topic, then listen to the tape and note the terms used.

Then, record in your journal any insights gained as you were devising the model, such as questions prompted by doing it, or strategies you might need to employ for data-making or analysis. You may find, for example, that it alerts you to the need to include particular people in your sample, or that you need to explore a broader context. You might also find it useful to create nodes to reflect the concepts you identified in the process of creating your model.

Later you can review the model to see how far your thinking has moved in response to gathering and working with data. An archive (***static***) copy of a model can be made, leaving the original version available to continue working on.

CREATING A SIMPLE MODEL

You can build a model from new or existing project items. It is absolutely not necessary to have already created some nodes, but if you have, you can use them to help build the model.

- ▶ Click on **Models** in the Navigation Pane, then create and name a new model using the RMB menu in List View. An area for working will be created in Detail View.

✓ To create more working space, from the menus select **Window > Docked** and the Detail View will become a separate window which can be enlarged to fill the screen. For now, you can close the Groups pane also (View menu).

Populate your model by placing items in whatever position you choose:

- To build with new items, drag a shape onto the model area. Double-click to name it.
- If you already have nodes, from the RMB menu, choose to **Add Project Items.** Check **Free Nodes** to bring in all your free nodes, or click once to open and select particular nodes. Do *not* add associated items (in this case, the journal).
- Items can be resized, or the shape can be extended in one or the other direction.
- Move the shapes to where you want them, by dragging. Multiple selections can be moved at the same time so that their spatial relationships are preserved.

Add connectors to show links between shapes or nodes (Figure 2.4):

- Select the first item for the linked pair. Use Ctrl-click to select the second item. While hovering over one of the selected items, choose **RMB > Add Connector,** and choose the type of connector that best describes the relationship between the two items. If you create a one-way arrow that is pointing the wrong way, select it and use **RMB > Reverse Direction** to fix it.

To archive the model:

- In the Detail View, **RMB > Create As Static Model.** You will be asked to name the new model—this is the one which will be the archive (static) copy, indicated by a different icon . Being static means it is unable to be changed, and it will lose any live links with project items.

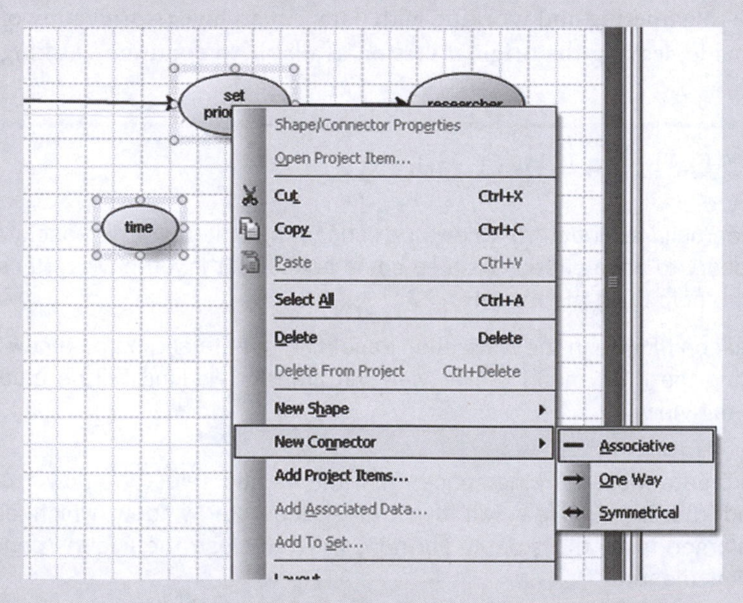

Figure 2.4 Adding a connector to a model

The starting model I drew for the Researchers project was clearly influenced by the years of experience I had already had working with academics new to a research oriented environment. What I realized from creating this model, however, was that motivation was not the end point in the process, other facilitative environmental and personal elements were still needed in order for people to be involved in researching.

SAVING YOUR PROJECT

As you have been working, NVivo has been regularly reminding you to save your work. There is no background autosaving in NVivo, so I would encourage you to respond to these reminders by clicking on **Yes** to ensure that your work is not lost. Of course, you should always save as you exit the project, as well.

For safety, you need backup copies, regularly updated, as a precaution against loss or drastic error. No matter how good your equipment, power failures or human intervention can make it crash; no matter how confident you are, errors can occur; and no matter how thorough the programmers and testers have been, occasionally projects will become corrupted. My recommendation is to make a backup on your working computer at the end of each day's work, and to copy that to another medium (a disk or memory stick, independent of that computer) on a regular basis. If you're cleaning out old files, a good housekeeping principle with backups is to keep the last known good copy – that means at least the last two backups, just in case the last one was corrupted as you saved it.

Backups can be made *after* you close your project by going into Windows (e.g., to your My Documents folder). Locate and copy the project, and paste it into a backup folder (see Backing Up in Help). So you can keep track of what is what in your backup folder, date your backup files (this also overcomes the problem of duplicate names). I use an international date format (yymmdd) added to the name, so that they sort correctly in a file list or navigation box (Figure 2.5).

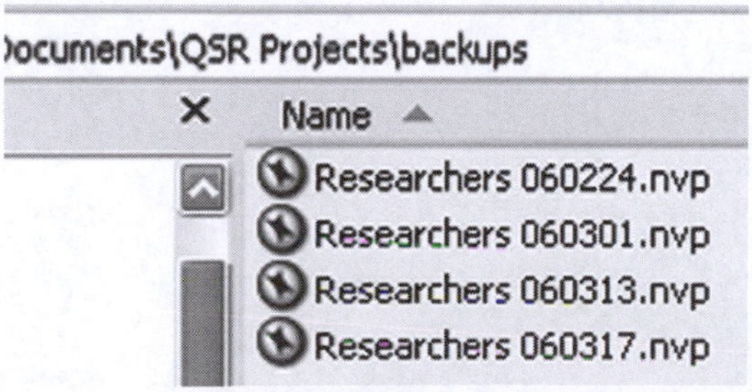

Figure 2.5 Storing backup copies of a project in Windows

NOTES

1 Except for setting Language options for those who do not work in English, these are suggestions only. They are provided primarily for those who become irritated by default options in new software. If you find them confusing, for now you can simply move on.
2 For those who are unfamiliar with the use of heading (and other) styles in Word, it is recommended that they take a little time to learn about use of styles for word processing (more on application of styles in Chapter 3).
3 NVivo now provides a Word Frequency Count through the Queries options. All finds for particular words can then be displayed and viewed in context.
4 If you are working in a situation where you are running a number of similar projects which use the same coding system (e.g., regular reviews of quality assurance in similar settings), it is possible to import a coding structure from one project into another (File > Import Project, and select to import just the Free/Tree Nodes, without coding).

Chapter 3

Making data records

It is possible to start working on your data really fast if you want to – just type up, save and import a document and you're ready to go. But you can gain significant advantages from careful preparation, especially if your data contain any regularities or structure that can be exploited to facilitate management or coding.

This chapter is full of hints and advice for preparing and managing your data sources – ignore it at the peril of wasting significant time later in your project! It is a chapter you need to read now, before you get too far in, and it is one you may need to come back to again, when you begin to work with different data.

This chapter:

- Alerts you to a wide range of data sources you might consider including in your project;
- Indicates the kind of details it is wise to record along with each source;
- Canvasses theoretical and practical issues that arise in transforming your research experiences, such as conversations with participants, into manageable sources for analysis;

Additionally, you will find practical guidelines on:

- determining what a case is, for the purpose of analysis in NVivo;
- how best to structure and arrange your documents, with variations related to methodological purposes;
- how to use heading styles to assist this. For those currently unfamiliar with use of Styles in Word, there is a fringe benefit: this can apply to and change your whole experience of using a word processor and preparing reports!

And finally:

- you will be introduced to some of the tools available for managing and reporting on document information.

While this chapter is primarily about making and managing data *records*, it is worth noting Coffey and Atkinson's (1996: 11) warning against making data "in a spirit of careless rapture ... with little thought for research problems and research design." Such enthusiasm tends to lead to the predicament of having "all these data" and not knowing what to do with them. In similar vein to Maxwell (2005), they emphasize the necessity for an overall methodological approach which will link questions, design, data and analytic approaches, and also that analysis is ongoing throughout the life of the project.

DATA FOR YOUR PROJECT

Different methodologies require different data and different ways of working with those data, although for the most part, data used in an NVivo project will be text based – "the good stuff of social science" (Ryan & Bernard, 2000: 769).

You will be creating data for your project right from the time you start thinking about it. Already you are recording ideas about what it is you are doing. Record observations also, and make notes from casual conversations with colleagues and with those 'in the field'. These will supplement the specifically designed forms of data which will be made or gathered to inform the question/s you are asking.

Most often, qualitative researchers think about working with transcribed records of interviews or focus group discussions, but not all interviews can be recorded, and not all data requires full transcription. Nor are data sources limited to freshly generated texts. Consider using comments added at the end of written questionnaires; records of observations; existing material such as media releases, nursing notes, or web material; or secondary sources obtained from qualitative data archives (Corti & Thompson, 2004). While NVivo is particularly appropriate for analysis of free-flowing texts, it is certainly not limited to that form of data; data in tabular form can be used, and pictures can be included within documents. (Convert graphs or other diagrams which are in formats that cannot be included in a document into pictures.) For data items that cannot be imported, such as archaeological artefacts or national park sites, summaries and coding information can be stored in special sources referred to in NVivo as ***externals***.

By recording and selecting materials to inform your project, you make them into 'data' (Richards, 2005). The problem for a project using qualitative data is not in generating data, which as Richards notes is "ridiculously easy" (2005: 33), but in making useful and manageable data records. Qualitative data records are often messy and large – an hour of interviewing, for example, can generate 25 pages or so of single-spaced text, and that text may course back and forth over a range of topics. Typically, data are complex – making them difficult to reduce, and data are contextualized – so that records of additional information about sources and perhaps also the settings in which they were obtained are needed also.

Including literature as background data

Beginning a project by reviewing what is already known on the subject of your research is a well established practice, as is reviewing the implications of relevant theories for your topic, and methods others have used to investigate it (Boote & Beile, 2005; Hart, 1999). Even in qualitative traditions where, in the past, an understanding was given that it was better to begin data collection without such prior investigation of the literature so as not to prejudice emerging understanding from informants or participants in the field, there has been a shift back to seeing value in viewing the literature as a source of stimulation and/or sensitization (e.g., Charmaz, 2006; Silverman, 2000; Strauss & Corbin, 1998) or perhaps as data for analysis in its own right (Caron & Bowers, 2000). In contrast, for example, a Foucauldian approach to discourse analysis assumes thorough exploration and analysis of historical archival material in order to understand current knowledges and practices (Kendall & Wickham, 2004).

In a practical vein, working with notes from literature or other archival sources in a project can be a good place to start developing skills and ideas using NVivo, as these are readily available while you finalize plans and gain the various approvals necessary for gathering your own data. Coding notes you have made while reading articles facilitates the development of a literature review, and will provide a basis for making comparisons with what you find from your own data. Don't delay until you have already written a literature review: one point of this exercise is to use NVivo to assist you in moving from rough notes to an organized review.

Whether you make contributions from the literature part of your main project, create a separate project for it, or alternatively, don't work with it in NVivo at all, is perhaps best determined on the basis of whether this body of literature will be used only for this particular project, or for others as well.[1] If you make a separate project for your review, but then decide you want to compare coding with your coding for other data, you can import the literature project into your main project and so bring the two projects together for that purpose (using ***sets*** to differentiate the different sources of data for analytic comparison).

Suggestions for efficient data preparation when handling reference material and notes recorded in a bibliographic program such as EndNote or ProCite are available on my website (www.researchsupport.com.au). Alternative suggestions for handling literature in NVivo are contained in a paper by Silvana de Gregorio, available at www.sdgassociates.com.

DATA IN CASES

Much of the structuring and management of data in NVivo is focused around the concept of a case. Use of the term *case* in this context is *not* intended to imply that you are necessarily undertaking case study research. Rather, cases are seen here as the *units of analysis* in a research study.

> In the sociological and anthropological literature, a case is typically regarded as a specific and bounded (in time and place) instance of a phenomenon selected for study. …Cases are generally characterized on the one hand by their concreteness and circumstantial specificity and on the other by their theoretical interest or generalizability. (Schwandt, 1997: 12)

Most studies will comprise several cases of a single type, such as a number of mothers, or patients, or sites, or perhaps decisions about care. The case may be someone around whom there is a constellation of people, as with a person with a disability where interviews have been conducted also with the person's caseworker, carer, and a family member. The main case unit(s) may have illustrative cases embedded within them, for example, a corporation may be the case, with one or more specific departments, or products, as illustrative cases within the study of that corporation (Yin, 2003). Or they may be layered, for example where schools, classes and the pupils in them are each treated as cases at different levels.

Much has been written about selection of cases in qualitative work. Patton (2002) gives a thorough overview of the range of sampling and selection possibilities, as do a number of other qualitative authors referred to throughout this book. Silverman (2000) also discusses issues in purposive or theoretical sampling, including those around the ability to generalize from a single case.

Yin (2003) warns to "beware" of cases which are not easily defined in terms of their boundaries – their beginning and end points. "As a general guide, your tentative definition of the unit of analysis (and therefore of the case) is related to the way you have defined your initial research questions," and, "Selection of the appropriate unit of analysis will occur when you accurately specify your primary research questions" (Yin, 2003: 23, 24). If your questions and conceptual framework are clear, it should take only minutes to define the case type (what kinds of units you are using) and thence the cases for the study (Miles & Huberman, 1994).

Cases in NVivo

Each new data record is optionally assigned to a ***case*** as you make or import it, or it can be assigned to a case at any time throughout the study. All data for each case is coded at a ***case node*** for that case (case nodes are accessed through the Nodes View). Attribute data for the case (e.g., demographics), also, are attached to case nodes. As you gather data, then, everything you know about a particular unit of analysis (a person, a site, a company, an event) can be brought together, can be given attribute values, and later, if appropriate, can be analyzed internally or compared with other cases.

A case node may code just one document, for example when each participant is interviewed just once, or a single case node may code several documents, as in the example above of the constellation of sources around the person with a disability, or when the same person is interviewed several times. Alternatively, the case node may code several parts of one or more documents, as with a focus group, field notes, or other types of multi-person documents, where data for each participant is held in (perhaps multiple) separated segments. Case nodes, then, can be particularly useful for bringing together 'chunks' of data from disparate sources, when those chunks of data all relate to a single case.

For an ethnographic study reviewing issues of research production and performance for academics in the arts, humanities and social sciences disciplines of a university (Bazeley, 2006), I created a case node for each member of academic staff, sorted by academic unit. Data for the study comprised administrative records of research funding received by each staff member and details of research publications produced by them (originally in two Excel spreadsheets), individually completed surveys, web profiles, media releases, field notes including both observations and interviews, other official records and incidental documentary sources. The case nodes brought together data from whole or part documents (mostly achieved through autocoding), so that I could instantly access everything I knew about a particular academic. Additionally, once all the data were coded for issues raised, and for topics of research (for example), I could easily discover which academics were concerned about (or demonstrated) which issues, whether there was sufficient interest in any particular topic to create a research group, and who might be interested in being part of such a group, including details of what their contribution might be.

DATA PREPARATION

Data records in NVivo are held in ***documents*** which, along with ***memos*** and ***externals***, comprise the ***sources*** for your project. Documents contain the raw materials from which ideas will be built about what is going on in the project, and evidence gathered to support growing conclusions. Having the documents be as 'true' to the original sources as possible is therefore important. Typically each document is a record of data-making with a particular person (such as an interview, or one person's open-ended survey responses) or a group of people (a focus group), or is focused on a particular situation (e.g., observations of the oncology ward for 16th May), or is of a particular kind (e.g., a batch of news reports on the upcoming election from the Tribune for April–June, 2006; or a set of company records).

A document is given a title, a description may be recorded, and it has content (usually just text, but perhaps including text in tables, or pictures). Formatting of text will be preserved on import, so you can make use of heading styles and other stylistic features (see below) to enhance the text and facilitate coding. Text can be saved for import, using Microsoft Word save options, in document format (.doc), rich text format (.rtf), or in plain text (.txt) – although this last option will lose all formatting features. Note that text which contains embedded objects cannot be imported: check **Help > Importing Sources** for more detailed information about the kinds of objects which will prevent import, and features which might be lost on import.[2]

Demographic and other categorical, numeric or date information relating to the source of each case should be stored safely alongside data records for later inclusion as ***attributes*** of cases.

Files containing on-line data which cannot be imported can be accessed via hyperlink from text within an NVivo document, as can web sites. (Hyperlinks are created *after* importing the document.)

Many forms of data require a minimum of special preparation (although they may still benefit from being given some), but the researcher with documents involving responses from multiple sources (including multiple speakers in a focus group), or documents which are structured in any regular or particular way, will benefit enormously from spending time structuring their data with heading styles. As well as focus groups, these data might include responses to standardized questions, observations marked by time or place, internet conversations, or multiple items of, say, news or internet messages. These kinds of data records can be formatted to capitalize on NVivo's tools for capturing and coding (i.e., autocoding) repetitive or structured headings.

Making transcriptions ...

> Transcribing involves translating from an oral language, with its own set of rules, to a written language with another set of rules. Transcripts are not copies or representation of some original reality, they are interpretative constructions that are useful tools for given purposes. Transcripts are decontextualized conversations ... a living, ongoing conversation is frozen into a written text. The words of the conversation, fleeting as the steps of an improvised dance, are fixated into static written words .... (Kvale, 1996: 165–7)

What appears at first sight as a purely mechanical task is, in fact, plagued with interpretive difficulties. Initially, this is when you discover the real value of using a high quality recorder for your interviews or, even more so, for group discussions (using two is recommended for the latter). There is also real value in doing your own transcribing, if at all possible – building knowledge of your data through what Frost and Stablein (1992) referred to as "handling your own rat" (one of their seven characteristics of exemplary research). At the very least, if another person typed the transcripts, it is absolutely essential for the person who did the interview to review and edit the transcript while listening carefully to the recording. A typist who unintentionally reorders or omits words can totally reverse the intended meaning in some sentences.

The flat form of the written words loses the emotional overtones and nuances of the spoken text, and so, as well as corrections, it is beneficial for the interviewer to format or annotate the text to assist in communicating what actually occurred with a view to the purpose and the intended audience for the transcription. "Transcription from tape to text involves a series of technical and interpretational issues for which, again, there are few standard rules but rather a series of choices to be made" (Kvale, 1996: 169). The goal in transcribing is to be as true to the conversation as possible, yet pragmatic in dealing with the data. Some issues and suggestions follow:

- A full transcript will include all ums, mmms, repetitions and the like. Repetitions communicate something about the thinking or emotion of the interviewee. Watch for typists who think they are helping by 'tidying up' the text! Repeated denials may in fact indicate an opposite meaning of what the respondent at first appeared to be saying (Kvale, 1996). 'Ums' may indicate hesitation or some other concern about the topic being discussed or event being recalled, although if they are simply a regularly occurring pattern of speech, the decision may be different.
- In the same vein, don't correct incomplete sentences (which tend to represent the way people talk) or poor grammar: it is important to capture the form and style of the participant's expression.
- Note events which create interruptions to the flow of the interview, for example: (tape off), (telephone rings). Note also other things that happen which may influence interpretation of the text (mother enters room).
- Record nonverbal and emotional elements of the conversation, such as (pause), (long pause), (laughter), (very emotional at this point). For some purposes, such as various forms of sociolinguistic analysis, even more detail may be required (e.g., the exact length of the pause, overlaps in talking). Emotional tone and use of rhetoric are important to record. For example, something said sarcastically, if simply recorded verbatim, may convey the opposite of the meaning intended.
- If one of the speakers (or the interviewer) is providing, say, a non-intrusive affirmation of what another is saying, one option is to record that affirmation simply by placing it in parentheses or square brackets within the flow of text [Int: mmm], rather than taking a new paragraph and unnecessarily breaking up the text flow.
- Digressions from the topic of the interview are a controversial issue. The decision about whether or not to include that text centres on whether there is any meaning in the digression. Unless there clearly is significance (or might be) in what was said, it is usually sufficient to skip the detail of that part of the conversation, and just record that there was a discussion about gardening, or comparing problems in bringing up boys (when the topic of the study was something quite different), and perhaps, how long it went on for. In practical terms, whether or not these digressions are included in full will often depend on who is doing the transcribing –

rarely will a disinterested typist be allowed responsibility for omitting a portion of the text.
- As you are listening to a tape and taking notes or transcribing, you may come to some point in the conversation that is particularly interesting or potent, where emotions are running high, or to an amusing exchange. Make a note in the document as to where that interesting bit is located in the original source (count 340). This will facilitate linking directly from your document to an external file with the tape extract, should you wish to be reminded of the tenor of that exchange.
- Translation, if required, creates particular problems for transcription, for example when the original language is needed to convey nuance of the text. The original words used can be inserted in parentheses to benefit a reader who understands the language (NVivo can handle text of multiple languages in the same document), or for a multi-lingual team using a single common language, an ***annotation*** might be used to describe more fully what was being communicated through those words.
- Some researchers like to comment on (annotate) the text as they are transcribing it. If you do this, then enclose your comment in unique markers to set it off from the transcribed text, e.g., <<comment>>. This kind of marker will allow you to see clearly that it is your comment rather than its being part of the conversation.

Kvale (1996) and Mishler (1991) provide useful discussions and examples of these and other issues involved in transcription. Whatever procedural decisions are made, they should be recorded, firstly for the typist(s), to ensure consistency in the transcription process, and then for the reader, to aid interpretation from the data.

Provide clear, written decision and formatting guidelines to your typist. You might also ask your typist to record her reflections on what she was hearing in the tapes as she transcribed them.

... or not transcribing

Interviews and other sources for sociolinguistic, phenomenological or psychological analysis generally should be fully transcribed. When nuances of expression are not needed for the analytic purpose of the research (e.g., when what is required from the research is to extract information about how something was done, or a list of relevant issues or indicators), verbatim transcriptions may not be needed; notes from those interviews may be adequate for the task. Using a computer program to assist analysis does not automatically mean you are required to use full transcripts.

When I had assistants interviewing researchers about the impact of receiving a small grant on their development as researchers, the conversations were taped. Much of the recorded conversation was not directly relevant to the topic of my research, however (researchers have an irrepressible urge always to tell what their research is about), and so the interviewers made notes from the tapes, supplemented by occasional quotes. Again, in a different setting, when Australia's corporate watchdog set out to investigate 'boiler room' share-selling scams, the researchers worked entirely from notes of their telephone conversations with those who had been approached by the scammers. These were sufficient to trace the chain of events and their responses to those as told by the interviewees (Australian Securities and Investments Commission, 2002) – deep emotional responses or phenomenological essences were not the concern of the sponsoring body!

Then there are always the times when you discover the tape recorder wasn't working … .

And the times when your participant opens up just as you have your 'hand on the door' to leave … .

When preparing notes from interviews, keep the comments in their spoken context rather than rearranging their content (e.g., as points under pre-determined topics).

Working from notes has the added advantage of a much reduced body of data to work through, although the data can be much more dense (as was the case in the boiler room study), making working with that body of data somewhat more difficult in the sense that coding may be attached to smaller and less contextualized units of data.

Formatting documents

Text in an NVivo document can incorporate most of the familiar richness of appearance that word processors provide, such as changes in font type, size and style, colour, spacing, justification, indents and tabs. So when you're making notes from or transcribing your interview, focus group or field observations, make use of this to help shape your data, express emphasis, convey the subtleties of what is happening, clarify how your respondents were expressing themselves, or to draw attention to critical statements.[3]

While there are no absolute formatting requirements for documents imported into NVivo, the addition of styled headings to break the documents into labelled passages can add value in many circumstances, and use of a table format to break the text into meaning units may be useful, for example, in some phenomenological traditions.

Using headings to structure the document

Use headings formatted with ***heading styles*** throughout the text to clarify its presentation, and to provide structure and organization to the document. The use of styles for headings means that a prescribed (built-in) style has been selected and applied to a heading. NVivo will not see text as a defined heading just because it is bold or in a different font.

As in Word, heading styles in NVivo are hierarchical, breaking a document into parts, with sub-parts (up to a maximum of nine levels). A heading level will include any text following to the point where there is another heading of the same level or higher. In a structured questionnaire or interview, such as is common in web surveys or evaluation studies, headings would be used to indicate the topic being discussed or question answered. The response to each question is then separated in the document. In a structured document, consistently labelling questions using a heading style means that NVivo's autocoding tool (*cf.* Chapter 4) can be used to find and code all the responses to each question.

In a less structured interview or focus group, speaker names might be set out as headings, so that turns in conversation are separated in the document. I usually recommend using Heading 2 for speaker names, so that Heading 1 is available should you choose to insert topic headings. Consistency in the level of heading chosen for a particular kind of item (such as speakers, or questions, or anything else you are using headings for) across *all* your documents is important, as this will facilitate use of autocoding and query tools in NVivo. Using headings for speaker names is not critical for interviews where there is just one participant, but (unless the group is homogeneous and identification of individual speakers is not needed) it does become critical in multi-person documents such as partner or group interviews where headings will allow the text to be identified or sorted by speaker. This makes it possible for you to store individual demographic information for each speaker.

When you are working in a node, the context of any retrieval can be easily obtained. A view of the heading-level context from which the retrieval came can give rapid access to information such as the topic being discussed at the time, who was speaking, or the question being answered. In field notes, viewing to the heading level might be used to check the date or site or circumstances of the events being described, while notes from literature can be structured so that the heading provides immediate access to the author and date for the segment being quoted.

- Have an observer present at your focus group, to record a speaker identifier and the first word or two said for each speaking turn. If the typist is having trouble identifying different voices from the tape, this can be used to verify who was speaking.

- The heading style needs to be 'pure and unadulterated' for NVivo to recognize it, that is, if Word (XP or later) has described a heading as, say, Heading 2 + Red (because you added the red after it was formatted as a heading), then NVivo won't properly recognize it as Heading 2. If you want your Heading 2 style to be red, then modify the actual style specifications to include the colour, or, if it is for a single instance of that type of heading, apply the colour *after* it has been imported into NVivo.

Tables in documents

Documents might have been prepared in tabular form, or tables might be included within documents. NVivo can handle either situation. In addition, tables can be created and/or modified (e.g., number of rows, number and size of columns) within the software. There are two situations in particular where the use (or management) of tables is particularly relevant.

Setting out data in tabular form may be useful for those working within a phenomenological tradition (or others with similar requirements for segmented text). Phenomenologists often recommend (after reviewing the text as a whole) that the text be broken into meaning units – passages defined by transitions in meaning, marked initially by slashes in the text. These can then be converted into tabular form, with additional columns added for interpretive comment, theme identification, or synthesis. Once the document is imported, text within cells in tables can be coded in the normal way, on a character or whole cell basis, and if appropriate, whole rows can be selected for coding (but not whole columns).

Those who have survey data in a spreadsheet or database need to extract the text responses to the open-ended questions for use in NVivo, to create formatted Word documents that are suitable for autocoding both cases and questions. These are much more appropriate for use in NVivo than the original table. Apart from the difficulty of fitting spreadsheet data into a page-based display, autocoding cannot be applied to a table within an NVivo document.

Line or paragraph numbering

Line or paragraph numbering is usually required for those working in discourse or conversation analysis, so that the exact location of particular features in the text can be specified. Those working in teams also find it useful to be able to compare notes by referring to a paragraph number in a transcript.

Automatic line (or paragraph) numbering applied in Word (**Page Setup > Layout > Line Numbers**) is lost on import, and (at the time of writing) NVivo does not show line or paragraph numbers on screen. If numbers are required, import the document into NVivo, export and save it with paragraph numbers (In List View, **RMB > Export Document**), delete the first import, and then tidy up and import the exported version. Multiple documents can be selected for exporting (or deleting) at one time.

Using Word to prepare documents

Detailed guidelines and instructions for efficiently using Word to prepare your documents (including for data in tables) are available on my website, at www.researchsupport.com.au.

Each Document folder in the Researchers project stores a different type of document. Explore these to see how headings have been used for interviews, focus groups, notes from literature, surveys and also to see how tables have been used to break up student experience data.

Naming and storing documents ready to import

Documents should be saved with a file name that can be used as a document name in NVivo. NVivo will sort documents alphabetically – it can be convenient to recognize this from the start, for example by starting the name for repeated interviews always the same way so that they are then placed adjacent to each other in lists.

It is generally best to keep the document name reasonably brief and straightforward. Don't include unnecessary words in the title, such as "Transcript of interview with ..." or dates or demographic details or even details of the context of the interview, as the name is then likely to be truncated in the display and they may all end up looking the same! File information such as location, interviewer and date of interview is more appropriately recorded (briefly!) in the first paragraph of the text, as this can be converted to a ***document description*** when the document is being imported.

Contextual details and associated field notes will be recorded in a memo for the document, while demographic and related details will be recorded as attributes of the cases associated with the documents. Document descriptions, memos and lists of attributes for the cases represented in documents can be accessed easily at any time and also can be printed in reports for reference purposes, but in a document title they are 'clutter'.

Documents prepared in Word for importing into NVivo can be stored anywhere in your computer system, though you will find it useful to have them all together in a clearly identified folder.

DATA RECORDS IN NVIVO

For most qualitative methods, data collection and analysis occurs recursively (Miles & Huberman, 1994) and so you will introduce data to a project gradually, as you analyze early documents and make decisions about what next is needed. If you are using questionnaire data or documentary sources, or if you made your data before you accessed the software, multiple documents can be imported at the same time. In that situation, the volume of data available from the start can be rather overwhelming and it is critically important that you plan its management. In any case, it is best to plan and set up a data management system before you start making data (Richards, 2005: 50–1), to consider what will be recorded, and how; where the records will be stored; and what additional information will be stored with them.

Creating documents

Record documents directly in NVivo, or prepare them in Word and import them into NVivo. For external sources which are not able to be imported, create a document to 'stand in' for that source. Each approach has advantages and disadvantages, and your choice will depend on your situation and preferences.

Recording directly in NVivo

When you created a journal for your project, you created a document within NVivo (in that case, as a memo). You can use the same process to create documents which might hold other forms of data, such as for ongoing field notes.

The advantage of recording field notes directly into NVivo is the easy ability to edit and add to the data records as you make further observations. Thus, if you are organizing your data in a way that requires regular (or irregular) additions to existing documents, it makes sense to do so within NVivo rather than writing elsewhere and then having to copy and paste into the existing document. Additionally, you can memo and/or code while making the record, which is particularly relevant where the record is created over a span of time. If, however, you are making a field record that you intend storing as a separate document, and are doing so in a single recording session, then you may wish to take advantage of the features of Word to do so, and import in the normal way.

In the university based ethnographic project referred to above, I kept a running record of field notes gathered over a 12 month period, comprising observations, notes from brief interviews and casual conversations, and notes from minor documentary sources pertaining to particular academics or academic units. These were in seven documents, one for each of the seven academic units included in my study. Each entry was prefaced by a heading to identify and segment it within that unit's document, each was dated. While it was still fresh in my mind, each entry was coded for content and to a case node for the relevant academic.

CREATING A DOCUMENT OR MEMO WITHIN NVIVO

- In the Navigation View for Sources, select the folder in which the document is to be created.
- Right-click in the List View (in the white space). Select **New Document** or **New Memo** (as appropriate), OR, click in the List View, and press **Ctrl+Shift+A** on your keyboard.
- A **Properties** dialogue will open. Type in a **Name** and **Description** for the document or memo. The new document will then open in Detail View.

Importing a document

Unless you're engaging in field work in which you are regularly adding to existing documents, the chances are that you will be importing the majority of your data files from Word, rather than creating them within NVivo. The advantages of creating documents outside NVivo and then importing are that you have access to a spelling checker, application of heading styles is more efficient, more advanced find and replace functions are available, and you will have a backup copy of the original document should you accidentally delete part of the text in NVivo (which, incidentally, is very easy to do). Additionally, if someone else is typing transcripts, they will not need a copy of the NVivo software. When a document is imported into NVivo, it is *copied* into the NVivo database, while the original document remains intact in its original location.

IMPORTING DOCUMENTS (OR MEMOS) INTO NVIVO

Documents saved as Word files (document format, rich text format, or if you're really desperate, plain text format) are easily imported into NVivo. As you import a document, there are a number of decisions you need to make about descriptions, cases, and text styles.

- In the Navigation View for Sources, select the folder into which the document is to be imported.
- Right-click in the List View. Select **Import Documents**. An **Import Documents** dialogue will open (Figure 3.1):
  - Navigate to locate the files you wish to import. Multiple documents can be selected for import using Shift+Click or Ctrl+Click.
  - Select **Update text styles** to change styles in the document being imported to match those you have set for the NVivo application (*cf.* Chapter 2) or for this project (in File > Project Properties); otherwise they will import as defined in Word.
  - Indicate if the first paragraph of the document/s should be used to **Create descriptions** for those documents.
  - If each document represents a case, then choose to **Code sources at cases** and indicate where they are to be located (probably directly under Cases, but possibly under a case group node within Cases).

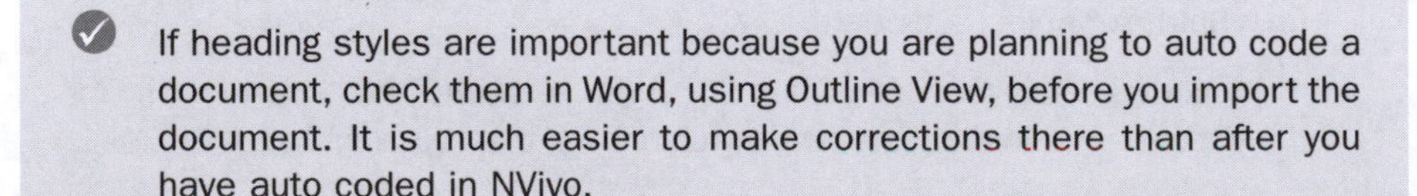
If heading styles are important because you are planning to auto code a document, check them in Word, using Outline View, before you import the document. It is much easier to make corrections there than after you have auto coded in NVivo.

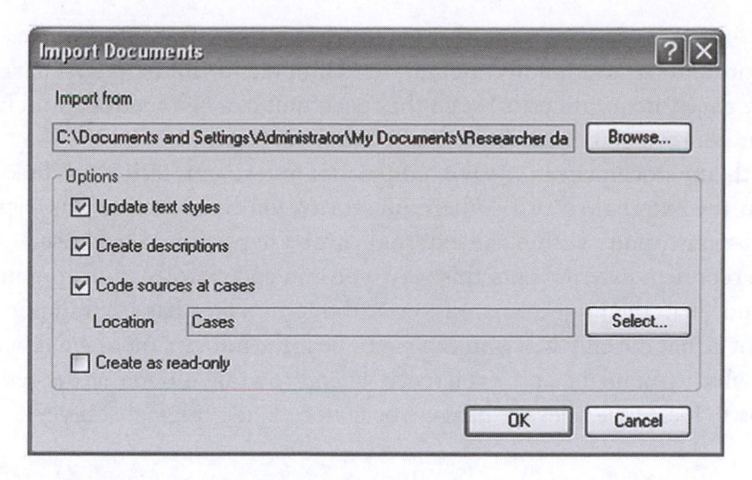

Figure 3.1 Import options for new documents

- Excel documents cannot be imported as documents into NVivo. If they contain text responses, convert them to Word documents; if they contain demographic or similar information that applies to cases, this data can be imported separately as attributes of the cases (*cf.* Chapter 6).

- A multi-case document (such as a focus group or survey data where the individuals are identified with headings) should not be assigned to a case when importing, rather, data for individuals within this will be coded to case nodes at a later stage using the autocoding tool.

- If your heading styles do not update correctly on import, this is an indication that there could be a problem in your original documents. Modify the original document to fix this and then re-import it before you start coding, as it will affect all functions which rely on the application of heading styles (such as viewing context and autocoding).

External sources

You might choose to include sources in the project as ***externals*** when:

- documents such as diaries or early research journals are not available on line;
- they are too large or are otherwise unsuitable for importing (such as a reference work, a thesis, artworks, music, videotape, photographs, maps, company reports, or a web site); or
- having access to the detail of the whole is not needed (perhaps for minutes of meetings or a set of guidelines).

A record is created within the program to indicate the existence of the document and to optionally define its component units (e.g., paragraphs, sections, pages, items, minutes) such that each unit can be identified and independently or consecutively coded, as you would code text.

As with any document in NVivo, additional notes, annotations or links can be added to the external record. Where illustrative on-line files such as pictures or videoclips exist, units within the external can be hyperlinked to those.

If you represent external data this way, you can code information contained in it and include it in the project analysis, even though no text has been imported. The constraint is that the only way you can see or hear the full version of what was coded (and to which you might find a reference when browsing a node or the results of a query), is to locate the relevant passage or feature in the original source.

For reference works and similar sources, my generally preferred alternative to creating an external of this type is to simply create (or import) a regular document in NVivo which contains a summary or description of the original source, with reference details (such as page numbers, grid references) where appropriate, to identify the material being described. Relevant information is then more accessible for coding, browsing or searching.

Access **Help** for further guidance about creating and using documents, including externals. Topics to view are: **About Sources** and follow the links for **Your Research in NVivo** and **Using the Software.**

Recording descriptive information about cases

Descriptive (demographic or other categorized) information which classifies your participants/cases can be recorded at any time prior to final analyses. This kind of information, which will be present in almost every project, is best stored using ***attributes***. If the information has to be extracted from your documents as you work through them, then it will make sense to record relevant attribute values as you are reading and/or coding each document. If, however, you have the information already stored in another form (as check lists gathered at the time of interview, or as survey data already in a spreadsheet or statistical database) you can choose to enter this information into your NVivo database at any time. This means that you might delay until you need it (which is generally not until you have imported and worked through a number of documents), as it is generally faster to enter it in a batch.

Attribute information will be used primarily to make comparisons across the data, for example, to compare male with female responses, or those of the experienced worker with those of the new recruit. How to create, manage and use attributes for your cases will be covered in detail in Chapter 6. For now you simply need to ensure the information is safely stored somewhere (on or off the computer), so you can find it when you need it.

Do *not* import attribute data which is in table format as a regular document, or even include it in a table as part of a text document, as you will be unable to use this effectively (or even at all) in that form. Attribute data set up in table format will be imported as a ***Casebook***, usually after your cases have been created.

Recording a document memo

In the previous chapter I introduced the routine of journaling as a way of keeping an audit trail of reflections on the project, spontaneous thoughts and developing ideas relating to the topic of the research as a whole. With the introduction of data documents to the project, it is time to consider another type of reflective document often used when working in NVivo – the document memo.

Each data file in NVivo can optionally be assigned one primary memo, that is, a linked document for recording additional observations, reflections and other materials relating to that item of data. Linked memos are available for nodes, also. You might prefer to attach your thoughts to the case rather than to document/s within it, by creating the memo from the case node.

At this stage, the memo might include:

- Field notes made following data collection, such as un-taped comments, observations, impressions;
- Thoughts about the meaning or significance of things that were said or written in this particular document;
- Key points and issues arising from the data to follow up in further data gathering.

You will, of course, be adding to this memo as you work in detail through the document (*cf*. Chapter 4).

If your thought or insight has significance beyond this particular document, it may be more appropriate to record it in the project journal, keeping the document's linked memo as a place for ideas arising specifically from this document. Whether you decide to create a memo for each document, each case, or simply use one general journal (in which you are careful to reference the sources prompting your recorded thoughts) will be a matter of methodological choice and/or pragmatic decision making, and may well vary from project to project. To avoid anxiety about losing track of where you have recorded the ideas and thoughts you have, I recommend that you code the content of the notes you make (preferably, as you make them). This will ensure the ideas always turn up whenever you consider those topics in your project.

CREATING A LINKED MEMO FOR A DOCUMENT OR CASE

- Select the document or case in List or Detail View.
- Choose **RMB > Memo Link > Link to New Memo** (or **Ctrl+Shift+K**). You can then define the properties of the new memo, and an icon indicating that a memo exists for the document or case will be evident in the List View for Sources or Nodes.

Once created, memos can be accessed when hovering over the document or case in either List or Detail View:

- Select **RMB > Memo Link > Open Linked Memo** (or **Ctrl+Shift+M**).
- Enter the time and date (**Ctrl+Shift+T**) and record and code text in the memo as you would any document.

MANAGING DATA SOURCES IN NVIVO

NVivo provides a range of tools to assist in data management, some of which are appropriately used from the beginning while others will become more relevant as the project develops.

Documents can be located within a user-defined ***folder*** in the Sources area. Folders might be used, for example, to store records of interviews separately from focus groups or survey responses, or perhaps to separate Phase 1 interviews from Phase 2 interviews, or Company A data from Company B data. As in any filing or computer system, you can create folders as needed to manage the kinds of data you are assembling in whatever way you choose. Documents can be placed in one folder only, however.

As indicated earlier, data about any cases are typically coded to *Cases* in the Nodes area, so that everything known about a particular case, and just that case, can be brought together. Attributes relating to the case are then stored with the case information. Text, whether from whole or part documents, would typically be coded at only one case node. One document, however, may contain text for several cases (e.g., a focus group), and a case node may reference (i.e., code) text from multiple documents (e.g., repeated interviews).

Finally, data sources can be organized in ***sets***, which primarily indicate that these data items belong together in some way. Sources can be in more than one set. Apart from providing a visual reminder of some common feature, sets of documents are used primarily when setting up queries.

Using folders, cases, attributes and sets to manage data will be discussed more extensively in Chapter 6 (*Managing data*). For now, the most important management tools to be aware of, and about which decisions may be needed as you are creating and importing documents, are document folders and cases (and even they can be dealt with later if you prefer).

Reviewing and arranging your sources

By now you may have several documents in your project. Check where you are up to, by looking at Sources in the Navigation View.

If you have several types of documents, or documents from different sources, and they are not arranged in folders, you might like to set those up now. If all your documents are of the one type, then of course there is no need to set up your own system of folders.

Click on each folder, in turn, to review the contents of that folder. Information for each document in the folder will show in the List View. Documents with a memo will have an additional icon showing in the column next to their name. Documents can be auto-sorted by clicking on the top of a column in the List View, for example, if you want to find your most recently added documents, or all the documents with no coding (a single text reference probably indicates coding to a case node only). NVivo will apply this sort order to all document folders, and will remember the order next time the list view is opened.

ARRANGING SOURCES AND VIEWING TEXT

Creating and using folders

- In the Navigation View for Sources, right-click on **Documents** to create and name a **New Folder**. Drag documents from the List View to the appropriate folder.

Viewing document text

- Double-click on a document in List View for the text to be shown in the Detail View, below the list of documents. More than one document can be open at a time, but the text of only one will be visible at any one time.
- Select which open document is currently in view by clicking on its tab at the top of the detail view.
- Close a document by clicking the ☒ (top right in Detail View).
- Right-click on a particular document to view the **Document Properties** (or press **Ctrl+Shift+P**). If it is helpful, you can add or change a description for the document in this dialogue.

✓ If you open your project and panic because you can't see any of your documents listed, it is likely to be because you've made folders and placed them in there.

✓ If you are closing several documents in a row, pause between each.

Making a report of a document

For most purposes, coding and memoing of text is done directly on screen, for maximum efficiency and flexibility in those tasks. You may want a printed copy of a document to code while you're travelling to work, however, or to use as the basis for a discussion in a team meeting, or if the document was created in NVivo you may feel more secure if you have a printed copy as well. For those kinds of situations, a printed report of the text of a *selected* document (or documents) can be obtained most easily by choosing **File > Print** (or **RMB > Print Document/s**). Choose to include paragraph numbers on the printed report if you are planning to code off-line and then use the ***paragraph coder*** to enter the coding information (*cf.* Chapter 4).

Alternatively, you can create and save a copy of the documents in Word by selecting the document (or documents) in List View, then choosing **RMB > Export Document.** Select the features you want to include in your report. A separate Word file for each of the documents you have selected will be created in the My Documents area of your computer (or in whatever folder you have set as a default location for reports).

 If you live outside North America, check the page size before printing.

Two final notes

1 Preparing data records is one of those areas where "fools rush in ..." Careful editing and thoughtful structuring both reap rewards for the analyst dependent on a representation of reality that is contained within a written record.

2 Sources are added to a project throughout the life of the project. Wherever possible, it is important to start the analysis (coding and memoing) process before you complete your data collection. Only then are you able to assure yourself that the data you are gathering will answer your research questions.

NOTES

1 For a one-off brief literature review, I tend to go straight to working in Word, using headings and a document map as tools for organizing what I am learning from the notes I already have in my bibliographic database or from new articles I'm reading.

2 The FAQ section on the QSR web site (www.qsrinternational.com) will also have up-to-date information on objects likely to cause import problems. If you are having a problem with a document which will not import, and you cannot easily see why, try the following strategies (with each being a little more drastic than the last in terms of what formatting features are lost):

(a) Save the document as HTML (Web) format, then resave as .doc or .rtf.
(b) Open the document in Wordpad (Start > All Programs > Accessories) and save it there as .rtf.
(c) If all else fails: save the document as plain text (.txt), then resave as .doc or .rtf (and re-format your headings using Replace, if necessary).

3 Although these features can be added to the text once it is imported into NVivo, in most cases it is easier to add them while working in Word and is often best done while checking the transcript (i.e., with the tape playing).

Chapter 4

Working with data

Qualitative analysis is about working intensively with rich data. The tools provided by NVivo support the analyst in making use of multiple strategies concurrently – reading, reflecting, coding, annotating, memoing, discussing, linking, visualizing – with the results of those activities recorded in nodes, memos, journals and models. Each of these strategies is integrated in a process of learning from the data, and indeed, they work best when they are carried out as integrated activities. The process of thinking about a code prompts a memo, and similarly, the process of writing an annotation or a memo assists in clarifying what a code is about, or which codes are most appropriate.

In this chapter:

- Discover strategies for seeing and naming the concepts and categories embedded in your data;
- Read, reflect, link and store ideas in annotations, memos and nodes as you work through the first of your sources;
- Begin to visualize the association of ideas or pattern of events in a model that sums up what you have learned about a case.

At first it will feel that progress in working with the data is slow, but as you work the project will grow into a web of data, categories and thinking, illuminating your research question. As your ideas and categories develop, working with the data will become faster.

Be prepared to experiment as you work; you are not trapped by your early work with data. The tools in NVivo are flexible, allowing for changes in conceptualization and organization as the project develops. Moving nodes does not mean they lose their coding links. Editing the text of documents does not invalidate earlier coding on that text. As your knowledge about your data and your confidence in using NVivo each gain in sophistication, it is very likely that you will reconsider and reconfigure what you have been doing. What you have already done will not be lost, and the effort already applied will not be wasted.

If you lay a sound foundation with your first documents, then you will confidently move on to adopt further strategies for working with and advancing your thinking about data, as outlined in following chapters. Whatever path you take, it is important to see this early work with documents and coding as beginning analysis, and not simply as preparation for analysis.

GOALS FOR EARLY WORK WITH DATA

It is always helpful to have a sense of what you are trying to achieve at any stage. Here is a general picture of what you might seek in this early work with your sources:

- You need both distance and closeness to secure a rounded perspective on your data (*cf.* Chapter 1). After working through a document, you should be surprised and excited and informed by nuances in the text, but also able to stand back and see the whole, and where that whole fits in a larger whole. Case overviews in the form of reports, written summaries and visual models will assist in gaining distance, as will reading and memoing nodes separately from the documents. Detailed coding and associated memoing will take you closer to an intimate knowledge of both the case and the ideas you are working with.
- Strive, even from this early stage, to develop the concepts you will be working with – to go beyond descriptive labelling and to think about them independent of the source. Why is this information important? Where will these ideas take me? This will be reflected in the way you name nodes, and in the memos you write.
- Early work with text and concepts is about laying the foundation for identification of key themes in the data. Beware of jumping to conclusions too early, however. Constantly challenge your first ideas through drawing comparisons, purposively sampling diverse cases, or by reviewing what the literature says on the topic. The project journal is probably the best place to record these ideas at this stage.
- Right from the start, it is helpful to identify patterns in the data: you will notice them as you are coding, for example, when you consistently find that you are applying particular codes at the same time. These hunches should also go into the project journal (mark them in some way so they stand out for when you review later); then you are off the starting block early, for final analyses.

Selecting documents for beginning analysis

If you are just beginning to gather data, selecting a first document to work on is probably easy: you will have only one or two from which to choose. If, however, you have already completed a number of interviews or have a variety of data items to choose from, then you might:

- choose one which you remember as being 'typical' in some way, or which was contributed by someone who was representative in some way of a group or subgroup in the sample;
- choose one which seemed to be particularly interesting, or 'rich' in its detail.

The first document you handle can have a significant influence in determining the categories you create and the ideas you carry through the analysis, as it will sensitize you to watch for certain types of detail. When choosing a second data item, therefore, you will benefit from selecting one that contrasts in some important way with the first. In addition, you are likely to generate the majority of your categories during coding of your first few documents, so it is useful to maximize the potential for variety in concepts (or in their forms of expression) early in the process.

I chose Frank and Elizabeth as the first two documents to work through in detail, because, as academic researchers, they provided an absolute contrast in terms of career development. Elizabeth's path into a research career was characterized by digressions and serendipitous events, while Frank's path was direct and purposeful.

When I first worked through Elizabeth's document, the impact of her changing image of research was striking, and so I dutifully coded that and further documents for the way in which the speaker viewed research – only to find in later analyses that it had no particular significance for anyone else. It became more useful to see (and code) Elizabeth's re-visioning of research as a *turning point*, rather than focusing on her image of research. The nodes that dealt with images of research could then be dropped (or 'retired').

GAINING PERSPECTIVE ON THE TEXT

Reading and thinking, using memos

Your first (re)reading of a document should be rapid but purposeful, directed but not bound by your research questions. The idea is to get a sense of the whole, so that as you begin to identify specific points or issues in the data, you will see those in the context of that whole. Reading right through before you start coding is especially important if it is some time since you did the interview or if someone else did it for you, or if your recent work on it was piecemeal. Make notes about what you think this document is telling you, reflect on a word or phrase, record your ideas about a concept or theme, note issues for further investigation and hunches to check out (Miles & Huberman, 1994). It doesn't matter if the typing or the grammar is rough, as long as you get the ideas down (Charmaz, 2006). Take the opportunity to discuss this document with a colleague, as that is likely to strengthen your reflective thinking about the text and its interpretation – then add these thoughts to your notes.

Many people prefer to scribble on hard copy at this stage, on scrap paper, or in a notebook, but there is a real advantage in making these notes on the computer. Add what you have learned from this reading to any notes you have made already in the document memo (*cf*. Chapter 3). This avoids losing ideas as they fade from memory, or become submerged in the morass of data. *Code the memo's text as you write, for the topic or concept it is about.* Because memos can be searched and queried, later you will be able to find just your thoughts on a particular topic, that is, just what you have written in memos, together with or separately from participant data.

Individual document memos are not necessarily useful for all projects, however. For example, in a project analyzing 120 brief case summaries relating to professional malpractice claims, I found it more useful to simply record the key issue for each claim in a single combined 'issues' document, and to use a separate journal for reflecting on what I was learning from various cases and for noting common themes or issues to explore. Individual document (or case) memos would have been 'overkill' – as indeed they would be for projects that contain just a few open-ended responses to questionnaires. Others find it more useful to link memos to the concepts being recorded, rather than to the documents (e.g., Bringer *et al*., 2006). For any project dealing intensively with rich data for a small sample, with data supplemented by field notes or a methodology involving case analysis, however, the memo for each document (or case) becomes an invaluable asset as it ties together the different threads of data for the case.

- ✔ If case nodes better represent your participants than do documents (when each document does not equate to an individual participant), you might record additional notes and reflections on a participant's text in a memo attached to their case node, rather than in a document memo.
- ✔ Keep the memo open as you work through a document, so you can easily flick between views.

Reading and thinking, using tables

Placing data in a table format, with rows defined by 'meaning units', is a strategy sometimes employed by phenomenologists and others. Use the additional columns to record commentary on, interpretations of, or abstractions from the segments of text (Giorgi & Giorgi, 2003; Moustakas, 1994; Smith & Osborn, 2003). Using a table in this way assists in clarifying the structure of and response to the experience being reported by the participant. Coding (probably applied to whole table rows) may be facilitated through this process; coding additionally replaces the need to use a column to list themes identified within the text. Document linked memos may be used in addition for field notes, to maintain the gestalt and to assist in developing a patterned understanding of that person or situation (Hycner, 1999).

 You can record comments and interpretations on the text using the table-based format either in Word before you import the document, or in NVivo after importing it.

Annotating, marking and linking

As you read (or later, as you code) in NVivo, you might also ***annotate*** words or phrases in the text. Annotations in NVivo work rather like a comments field or a footnote in Word. Whereas the document memo or project journal is more useful for storing reflective thoughts and ideas from the text (discussions with yourself!), annotations are useful for jottings or notes about a particular segment of text. You might use annotations also to note the intonation of the voice at that point in the conversation, or a translation or transcription problem, or to comment on some aspect of the discourse. For example, when a coach referred to a celebrity in her study of sexual abuse in elite swimming, Joy Bringer used an annotation to note the significance of that reference to the conversation (Bringer *et al.*, 2006: 249). You can create annotations while working in either sources or nodes, and they will travel wherever their text anchor goes, which means that you can view them while you are reading the text in a node or if that text turns up in the results of a query. They work well, then, as reminders of things you need to be aware of whenever you read this bit of text.

If you are reading on screen and an interesting expression or detail draws your attention, change the text to a colour so that it stands out in retrievals or on later review. Highlight the text to be coloured, and select a colour from the formatting toolbar (next to the font size). Then write a note in a memo or annotation about *why* it appears significant.

- The text of annotations can be searched but not coded. Thus, if you want to find particular comments later, either code the 'anchor' for the annotation or include a relevant keyword in the comment.
- Use colour also to highlight questions arising in your memos, so that you are reminded later to check if you have found an answer.
- Be careful not to move the text when you select it. If you do accidentally move it, immediately click on Undo to correct it.

As you move into using coding tools to work with your document, continue to reflect on the meaning of the text and the concepts you are developing from it, especially during these early stages of the project. If you find you are focusing on a particular concept, rather than the case, you may prefer to create and link to a memo about that. This is best achieved using a ***see also link*** which connects from an anchor in the text (*cf.* Table 4.1). Create a new memo, named for the concept – later you can link this to an appropriate node – or open an existing one.

Occasionally, you might want to link a passage to an external file, a website, another document in your project, or to a particular passage in the same or another document (perhaps even the document's own memo). In NVivo, these are referred to as ***hyperlinks*** and ***see also links***. For example:

- *Hyperlink* to a non-project, on-line file (picture, sound track, report, .pdf article) that you can't, or don't want to import, or to a web page (assumes connection to the internet);
- Where the text illustrates something you have read in the literature, create a *see also link* from that text to the relevant material in a reference document (either a passage in a document containing notes from literature, or an external);
- Use the capacity to create a *see also link* from one passage to another in the same document to point up contradictions in a narrative;
- Show a sequence of events by creating a series of linked passages or documents, using *see also links*.

Use of see also links and node memos will be explored further in the following chapters. For now, so you can put things in perspective, Table 4.1 outlines the key features of various linked memos, annotations, see also links and hyperlinks, to help you gain a sense of when each might be most useful, and what are the limitations of each.

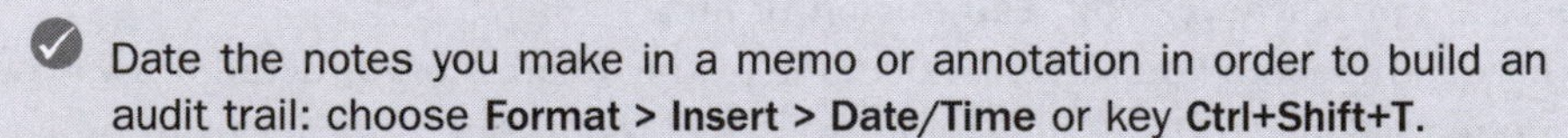

✓ Date the notes you make in a memo or annotation in order to build an audit trail: choose **Format > Insert > Date/Time** or key **Ctrl+Shift+T**.

✓ It is helpful in naming memos to make it clear if they have a primary relationship to a document or node. If you lose track of what is linked to what, or linked things unintentionally, then: (a) you can check what a memo is linked to by going to Links in the Navigation View; (b) you can delete a link without deleting the memo (RMB option); and (c) you can rename a memo to better reflect what it is about.

TABLE 4.1 Memos, annotations or links: which should it be?

	Document linked memos	Node linked memos	Annotations	See also links	Hyperlinks
Primary use	Field notes and observations; reflective thoughts about the source as a whole or points in it	Reflective thoughts about the concept or case represented by the node; ideas for further analysis	Notes which illuminate or reflect on a specific part of the text (seen in a document or node)	Links from a specific point in the text to project items of any kind, or to specific content in memos or other documents.	Links from points within documents or externals to non-project on-line items or websites
Display to indicate presence	Icon next to source in List View	Icon next to node in List View	Blue highlight on text	Red wavy line under text	Underlined blue text
Coding of linked item	Can be coded	Can be coded	Content can't be coded (code the anchor)	N/A (code the anchor)	Can't be coded (code the anchor)
Searching for a word or phrase	Content can be searched	Content can be searched	Content can be searched	N/A	Can't be searched
How many can you have?	One linked memo per document	One linked memo per node	As many as are needed	As many as are needed	As many as are needed
Help topic	About memo links	About memo links	About annotations	About see also links	Hyperlinks
To create:	From source item in either List or Detail View: RMB>Memo Link>Link to New Memo; or Ctrl+Shift+K		RMB>Links from selected text in source or node; or select icon (or) in Links toolbar		RMB>Links from selected text in source
To view:	RMB>Memo Link>Open Linked Memo; or key Ctrl+Shift+M		View>Annotations; or click on in the View toolbar	RMB>Links>Open to Item	Ctrl+click on highlighted text

BUILDING KNOWLEDGE OF THE DATA THROUGH CODING

Coding is one of several methods of working with and building knowledge about data; use it in conjunction with annotating, memoing, linking and modelling. "Any researcher who wishes to become proficient at doing qualitative analysis must learn to code well and easily. The excellence of the research rests in large part on the excellence of coding" (Strauss, 1987: 27). Guidelines outlining the basics of coding follow, but exactly how and what you code will vary significantly, depending on your choice of methodology.

Codes and coding

A code is an abstract representation of an object or phenomenon (Strauss & Corbin, 1998), or more prosaically, a mnemonic device used to identify themes in a text (Ryan & Bernard, 2000). Codes range from being purely descriptive – this event occurred in the *playground*, through labels for topics or themes – this is about *violence between children*, to more interpretive or analytical concepts – it is a reflection of *cultural stereotyping* (Richards, 2005).

Raw field notes and verbatim transcripts reflect "the undigested complexity of reality" (Patton, 2002: 463), needing classification to make sense of them, and to bring order out of chaos. Coding in qualitative research, in its simplest sense, is a way of classifying and then 'tagging' text with codes, or of indexing it, in order to facilitate later retrieval (Coffey & Atkinson, 1996; Miles & Huberman, 1994; Ryan & Bernard, 2000). Text then can be viewed by category as well as by source, and so, as well as facilitating data management, classification of text using codes assists conceptualization. This 'recontextualization' of the data (Tesch, 1990), through which data is seen afresh, assists the researcher to move from document analysis to theorizing.

Applying a code is often thought of as a reductionist process, as indeed it is when numeric codes are used to represent an experience, characteristic or attitude. In qualitative analysis, however, it is seen as a way of linking data to ideas and from ideas back to supporting data (Richards & Morse, 2007). This kind of linking facilitates data retention rather than data reduction, as access to the data (and the ideas) is retained (Richards, 2005).

As well as linking to data, codes link to each other. As you code, patterns of association between codes will become apparent. You gain value, therefore, not only through identification of relevant concepts but also in establishing and thinking about the linkages between them (Coffey & Atkinson, 1996). It is important to note potential linkages as you become aware of them (at this stage, in the project journal or in a memo attached to a document or node): these will feed into further theoretical thinking and analyses.

Approaches to coding

When it comes to the actual task of coding, there are "splitters" – those who maximize differences between text passages, looking for fine-grained themes, and "lumpers" – those who minimize them, looking for overarching themes (Ryan & Bernard, 2003: 95), and then there are those who 'have a bet each way' and do a little of each. A common approach is to start with some general categories, then code in more detail (e.g., Coffey & Atkinson, 1996), while those working using grounded theory, phenomenology or discourse analysis more often start with detailed analysis and work up to broader categories. If you are feeling a bit uncertain about how to tackle the coding task (most of us do when first faced with a complex paragraph of rich data), then a rough sorting of data into major categories may be a useful way of getting started – but note, you will need to go back and take a second look. Most end up working with some combination of the two approaches, and the software, happily, supports either or both.

Lynn Kemp, of the Centre for Health Equity Training Research and Evaluation at the University of New South Wales, tells students that choices about coding are like choices in sorting the wash. Some hang the clothes just as they come out of the basket, and when dry, throw them into piles for each person in the family, ready for further sorting and putting away – hopefully by their owner! Others hang the clothes in clusters according to their owner, so that they are already person-sorted as they come off the line, although pants and shirts may be mixed up and socks still need pairing. And yet others hang socks in pairs and all like things together, so that they can be folded together as they come off the line. Ultimately, the wash is sorted, people get their clothes and (hopefully) the socks all have pairs. Similarly, whether you start big, then attend to detail, or start small and then combine or group, your coding will eventually reach the level required.

Broad-brush or 'bucket' coding

This one is for lumpers! Because you can code from the text at a node just as you do from a document view, there is no need ever to treat coding as unchangeable – you can code on from already coded data (Richards, 2005). Your initial coding task, therefore, may simply be to 'chunk' the text into broad topic areas, as a first step to seeing what is there, or it may be that you want to identify just those passages which will be relevant to your investigation.

Sorting of text using this type of coding allows you to:

- See broadly what you have in the various areas you are covering to determine, for example, if there are some areas which will need more data;
- Identify text that is particularly relevant to the areas you need to focus on for now, where your interview questions prompted a broad ranging discussion but your write up has to be focused on just one part (see illustration below);
- Set aside, or 'park' text that you are not wanting to think about in detail just yet, or which will be useful for a subsequent study. Some of the interview passages for the Researchers study, for example, came from a single *motivation* node in an earlier, detailed study of early career researchers' experience with research funding agencies.

- Identify sequences from within the larger corpus which focus on a particular type of exchange (e.g., for those undertaking conversation or discursive analysis);
- Complete some preliminary analyses as a guide to where and how you will then proceed; for example, by sorting what was broadly said about a topic by members of different groups, you might determine that there appears to be enough variation to make this issue worth pursuing with more fine-grained coding or more data;
- Sort answers according to questions that were asked, particularly where they were gathered through a structured survey/questionnaire (*cf*: ***auto coding***, below);
- Code the contextual arena or circumstances which form the backdrop for the main issues, for example:
    - In a study of children's behaviour, code whether the behaviour being talked about is occurring in the home, the school playground, the classroom, or in the community;
    - In studying a process of change in a business, code whether the stage being talked about is pre-implementation, early implementation, later development, or post implementation. Then code in detail what actually occurred and the responses to it.

  This will facilitate querying the data later, to see if the behaviour or issue in question varies according to circumstance or timing.

Lynn Kemp's doctoral study of the community service needs of people with spinal injuries, which was the basis for the Get on with Living tutorial in NVivo2, employed broad-brush coding as an initial sorting strategy. In response to her question, "You came home from hospital and ...?" her interviewees talked extensively across all areas of their lives. In order to manage this large pool of data, Lynn coded her interviews first at very broad categories (e.g., community services, employment, education, recreation). She then coded on from the community services text (which was her immediate focus), capturing the detail of what it was that people with spinal injury were seeking from life, what services were offering, and how these supported or impeded their client's capacity to fulfil their 'plan of life'. After recovering from the doctoral process, Lynn (or indeed, her students) could then focus attention on topics that were set aside, to engage in further analysis and reporting.

> In the Researchers project, broad nodes code when the researchers were talking about *becoming a researcher*, and when they were *being a researcher*. Later, I used these nodes to see how stage of development was associated with different strategies or experiences for researchers.

Coding detail

And this one is for the splitters! While broad-brush coding relies on the capacity of the software to facilitate recoding of, or coding on from text at a node, coding in detail will make use of the capacity in the software to ***merge nodes***, to cluster like things together under a ***tree node***, or to gather related concepts in a ***set***.

For some methods, most notably grounded theory, initial analysis typically involves detailed, slow, reflective exploration of early texts – doing line-by-line coding, reading between the lines, identifying concepts and thinking about all of each concept's possible meanings as a way of 'breaking open' the text, to be recorded in both codes and memos. At the beginning of the analysis process, you explore each word or phrase for meaning, perhaps exploring theoretically the difference it might have made if an alternative word had been used or a contrasting situation described, or how this chosen word or phrase is similar to or different from others used (Strauss, 1987). Microanalysis of this type generates an awareness of the richness of the data, of how many ideas can be sparked by it and of how much can be learned from it, and a coding process that involves detailed attention to the text helps you to focus attention on the text rather than on your preconceptions. Detailed coding of this type also helps you see patterns in what participants are saying or documents are reporting. This leads you to identify the concept of which these instances are indicators and the dimensions of what is being described, that is, the ways in which that concept might vary.

In practical terms, capturing the detail of the text does not mean that you should segment it into tiny, meaningless chunks. Rather, the goal is to capture the finer nuances of meaning that lie within the text, coding enough in each instance to provide sufficient context, without clouding the integrity of the coded passage by inclusion of text with a different meaning (unless, of course, the whole passage contains contradictory messages).

Frank begins his response to a (grounded theory style) question about where he'd come from in research terms as follows:

> My PhD was both a theoretical and empirical piece of work; I was using techniques which were novel at the time. (interruption by secretary). I was using novel mathematical dynamic techniques and theory and also I was testing these models out econometrically on cross country data. I think it was a strong PhD, I had a strong supervisor, he was regarded as – it is fair to say he would have been in the top 5 in his area, say, in the world.

In the first place, it is interesting that Frank begins his response by focusing on his PhD experience. In broad terms, one could simply code this passage descriptively as being about the PhD or learning phase of becoming a researcher, and indeed, that may be relevant as contextual (or structural) coding even if one is looking to capture detail in the text. But this passage is also saying something considerably more than just that Frank's research career included a PhD student phase. He implies that the PhD provided a strong foundation for his later career. Linked with that strong foundation are the kind of work he did in his PhD and the role of his supervisor.

This text tells us also a great deal about Frank and his approach to research. His PhD was both theoretical and empirical (characteristics which are repeated in the next sentence). Not only does he have credibility in both these aspects of his discipline, but his work was novel. Here, he is both validating and emphasizing the strength of his foundational work. He is also suggesting, in describing his work as using novel mathematical techniques, that he is making a mark on the development of the discipline – presaging a later, critical theme for Frank.

That his PhD was a "piece of work" also suggests wholeness or completeness, so research work can have the dimension of being partial, incomplete and ongoing, or of being finished and completed with loose ends tidied up. This, along with the dimension of research discoveries as being novel vs incremental developments, may make for interesting thoughts about the nature of research activity (and being a researcher), but are not necessarily relevant to the current question of how one becomes a researcher.

Frank's emphasis on his PhD being strong and (noting repeat use of "strong") on the strength of his supervisor brings into more focus the idea also evident in the first sentence – that the status of his work matters. It is important to him that his supervisor was at the top, and that he was moving up there too (with leading-edge work). This theme of ambition (incorporating status) is evident also in several further passages, for example:

> I then teamed up quite soon with another 'young turk' ... we hit it off and both of us were interested in making some sort of an impact, so basically we were young and single and just really went for it. ... our two names became synonymous with this certain approach, and we went around Britain and into Europe giving research papers and getting ourselves published. ... it was us carving out a niche for ourselves in the research part of the profession and that was our key to success so we kept at it.

How important is this level of ambition, both from the point of view of the research question, and for detailed coding? At this early stage, the safe move is to create a node for ambition: if no one else talks in these terms, then later it may be combined within a broader category of, say, drive or commitment.

Coding for the opening passage might therefore look as shown in Figure 4.1. Other ideas and reflective thinking (noted above) prompted by the passage were recorded in annotations and in the memo attached to Frank's document:

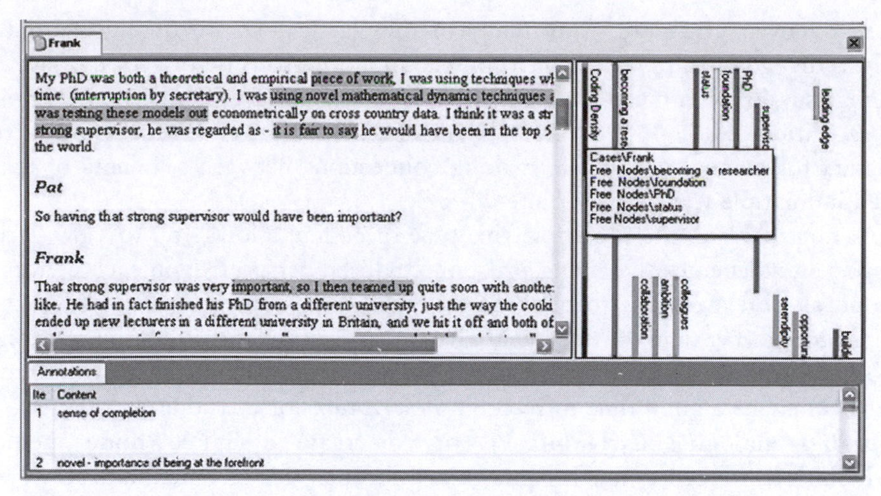

Figure 4.1 Frank's document showing annotations and coding

You might use annotations as a 'quick note' area while you are engaged in coding, and then, when you have finished going through the document, make a report of them and review them as a basis for more reflective memoing.

Fracturing or slicing data

The coding on Frank's document illustrates clearly that multiple nodes can be and are used in coding a passage. Coding is a way of 'fracturing' or 'slicing' the text,[1] of resolving data into its constituent components (Dey, 1993). Each passage is read carefully to identify the who, what, when, where and how of what is going on there, and each of these components is recorded as a separate code. Thus, multiple codes are used to capture what is happening in a single passage of text, but all that is known about a particular 'who' or 'how' from across multiple passages is held together in one code.

Strauss (1987) linked the fracturing process with use of his coding paradigm – asking questions about what conditions, interactions, strategies or consequences might relate to an initial code on the text, as a way of opening up the data and prompting valuable memos about (and recording of) related concepts. Using a computer for analysis makes doing this without losing the perspective of the whole uniquely possible, as the software can easily locate and reveal passages identified by intersecting codes, that is, it can recombine sliced passages that have been coded by 'this' and 'that' as well as 'the other' (*cf.* Chapter 5). The challenge to the novice user is to trust that those connections remain, and that you can reconstruct the links in the data – node to text, node to node – as needed. This is a step removed, therefore, from many of the classic texts on qualitative data analysis that give examples in which context and/or person and/or action or event and/or issue are combined in a single code (e.g., Miles & Huberman, 1994: 59–60; Patton, 2002: 464, 516–17). This is because those coding schemes were designed for manual use, where making connections between segments of text coded in multiple ways can be quite difficult.

At a first level of interpretation, this type of coding allows you to view each component independently, giving a recontextualized perspective on each concept or topic as all text relating to it is brought together. Seeing your data in terms of the category rather than the document gives a stronger sense of what the category is about (Richards, 2005). If this category is potentially significant in your project, this may be a good time to record a ***description*** for the node (in the node's ***properties*** dialogue: **Ctrl+Shift+P**), or to create a linked node memo (**Ctrl+Shift+K**) to record new insights gained through this recontextualized view.

At a second level of interpretation, slicing data into its component parts opens up analytical possibilities through the recombination of coded passages. Actions might be viewed in association with contexts, or issues with responses, to see the pattern of which ones are linked (because they were coded together on the same passage). Further, experiences or processes could be discovered to be somewhat different for members of different groups, or under different circumstances. This kind of recombination uses a ***coding query***, typically involving a pairing of several nodes to create a matrix (tabular) display. Through this type of query I discovered, for example, that researchers talked about *time management strategies* primarily in the context of their *being a researcher*, not when they were *becoming a researcher*.

Finally, slicing data into its component parts in this way will avoid repetitive nodes that result in data about a particular concept being located in several different places. For example, if a student talks about being encouraged by their supervisor, it is natural to want to create a node *encouragement by supervisor*, but if I do so, I am combining an action (encouragement) with a reference to a person (supervisor). Then I find that someone talks about being encouraged by his or her family, or by a friend, and someone else about being encouraged by a colleague, and so on. In such a system, I would need to create separate nodes for each combination of encouragement with source person, thus repeating *encouragement* several times, and having the text about encouragement split across several nodes. Additionally, some of these same people might be sources of discouragement or some other kind of interference to developing as a researcher (family, for example – or supervisor!), and so the nodes keep on multiplying, almost exponentially (T. Richards, 2004). Slicing data into its component parts by coding the same data into multiple nodes, rather than fracturing the category by splitting text about it across several nodes, avoids this problem. This means there needs to be just one node for encouragement, and one for discouragement, and one node only for supervisor, one for family, one for colleague, and so on – creating many less nodes in total. Each passage of text is coded (for this example) *as a whole* for what action is occurring *and* for who or what is the source of that action. This allows everything I know about encouragement (or discouragement) to be reviewed and recontextualized; and everything I know about encouragement or discouragement to be recombined with the source of that experience (using a coding query or matrix coding query, *cf.* Chapters 5 and 8) to discover patterns in who provided either of these. Furthermore, because all the data about encouragement is held together, it allows me to ask additional questions about encouragement, for example, about the stage at which it is important, the context in which it is provided, or what the consequences are of its provision in different circumstances.

Strategies for identifying and naming codes

The idea of coding a passage of text to index its content sounds simple enough, and observing someone else's coding can make the task look deceptively easy. When you meet your own text, however, you find that many things are going on at once: something is happening in a particular setting or at a particular time, particular people or groups are involved, perhaps their responses are based on their belief systems or cultural background, and there are consequences to be considered. Perhaps there is a particular twist to the way this experience, belief or feeling is being reported that makes it just a bit different from other reports, or difficult 'to get a handle on'. Narrative is inherently complex: a group can argue about the content and meaning of just one paragraph for a very long time.[2]

Resolving the narrative into its constituent components can help you deal with the complexity. Rather than trying to make a single code which captures the multiple aspects of a particular text, as you would if working on paper, code each component separately, that is, use multiple codes for the same passage of text – the computer will be able to reconstruct the links when needed. If you feel you are losing the overall picture in this process, write a memo about the complexity of the narrative, noting the connections that are evident. Later you can check whether these connections form a pattern, or were a 'one off'.

Identifying topics in a coding system has been described as being similar to the construction of an index for a book or labels for a file system (Kelle, 2004; Patton 2002).[3] But, more than simply labelling, naming a concept or topic aids classification of data and so assists analytic thinking (Dey, 1993; Strauss & Corbin, 1998).

What follows is a collection of ways in which codes might be identified and named, particularly those which comprise more than a simple descriptive label. In NVivo, a named concept or category or code is stored at a ***node*** (see below for more on making and using nodes). Node names are not fixed, so if you find the name you have given a node no longer fits well, change it by selecting it, clicking on it again and then editing, or by choosing to view its properties (Ctrl+Shift+P) and changing it in the properties dialogue.

As you create a node, get into the habit of documenting what it is. Name it carefully with a meaningful title, and if the name doesn't provide clear identification, add a description (through the properties dialogue). Descriptions serve as accessible reminders of what this node is about – and they are *very* useful when there are multiple people working on the same project, or when other responsibilities take you away from this project for a time, and you then have to pick it up again.

 If you are distracted by interesting but possibly irrelevant thoughts, and you want to save the thoughts, or the text, 'for later' (or just feel bad about throwing them away), create a 'bucket node' and/or a memo to drop them into. Then, when this project is finished and you want to return to those other ideas for another paper, you can quickly access them and code on in detail from that node.

'Seeing as': generating conceptual codes

> If sensing a pattern or "occurrence" can be called seeing, then the encoding of it can be called seeing as. That is, you first make the observation that something important or notable is occurring, and then you classify or describe it ... [T]he seeing as provides us with a link between a new or emergent pattern and any and all patterns that we have observed and considered previously. It also provides a link to any and all patterns that others have observed and considered previously through reading. (Boyatzis, 1998: 4; quoted in Patton, 2002: 463)

At first, you may not be sure quite what is relevant, or how it will be relevant. It is very easy, when coding, to be beguiled by fascinating things your participants have said and so to become sidetracked. To 'break in' to the text and make a start, try the following three steps that Lyn Richards developed for undergraduate teaching (Richards, 2005). They will help you move from 'seeing' to 'seeing as':

- Identify: What's interesting? Highlight the passage.
- Ask: *Why* is it interesting? This may generate a useful descriptive code or perhaps an interpretive code – if so, make a node for it. It may also warrant a comment in an annotation or memo.
- Then ask: Why am I interested in *that*? This will 'lift you off the page' to generate a more abstract and generally applicable concept, which, if relevant to your project, will be very worthy of a node (and perhaps a memo).

What that critical third question is giving you is a way of generating concepts for nodes that will be useful *across* documents, rather than nodes which code only one or two passages in a single document – these more general or abstract concepts are essential for moving from description to analysis (Strauss & Corbin, 1998). These nodes will link also to the broader field of knowledge, and it may be well worth recording node memos for them. In a strategic sense, that third question also helps keep you 'on target', to keep the focus on issues relevant to your research questions.

In a doctoral project on the role of support groups in assisting young people with a mental health problem, a participant reported moving from sheltered to independent accommodation. The student deemed this to be of interest, and created the node accommodation *to code it. Only one participant, however, had anything to say about accommodation, and accommodation was not an issue of concern for this project. (Had the project been about issues faced by young people with mental health problems, then of course, accommodation could well have been a relevant node to make.) When the student was challenged about why the text at that node was interesting, it became clear that what was of interest was the evidence of improvement in mental health status indicated by the shift in accommodation. She then changed the node name to reflect this more pertinent (and more interpretive) concept. She could then use that node across other documents to code other indicators spoken of in the same way, such as gaining employment or repairing a relationship. If she wished to examine the nature of the evidence given to indicate improvement in mental health status, a simple review of the text coded at the node would reveal that.*

Shane was asked if he could pin-point a time when research became of particular interest to him, or whether it was just something that was always there. His response: "Yes I can, with precision," followed by an elaboration of when and how, was interpreted by me simply as indicating a critical turning point experience – until I was engaged in a discussion with others around this sample of text. Their first thought, in contrast, was to see this as an interesting indicator of personality, the careful, attention-to-detailed way in which he thought about things – with no thought of turning points or critical experiences. In later text, he described how he went about the research task, exercising great care with details of history. His detailed work of identifying and piecing together archival snippets of information for historical analysis and reporting may help to explain why he, with this personality trait, was captured by it. So, at an early stage in the process of coding, as well as coding *turning point or critical experience* it would also be useful to code, descriptively, *attention to detail* as a personality characteristic, and more analytically, to code and/or note (memo) the *match between personality and research task*. It is probably better to be more rather than less inclusive early on, and then later, if any nodes appear not to be useful for further coding and analysis, they can be modified, combined or dropped.

A priori, or theoretically derived codes

Those working from a background of extensive reading in the literature, who have a lot of prior experience, or who are bringing a strong theoretical basis to their investigation will come to their data already with a start list of concepts they are interested in exploring, developing or testing with new data. It is useful, as a minimum, to analyze your research question and identify from within it every category or concept that is used, knowing that you will need a node for each concept in order to gather (and then relate) data about them.

While having a list of *a priori* codes can be useful (especially in a focused or time-limited study), it can confine your reading of the text, and so the advice is to 'hang loose', feel free to change or develop what you have set up, as you delve into the data. At the conclusion of their study, Meijer *et al.* noted:

> In the investigation of teachers' practical knowledge, it is important to let the data speak for itself as much as possible. Yet, there are insights available about teachers' practical knowledge which can be legitimately used in further investigations about this concept. In our study, we used insights generated from research on the content of teachers' practical knowledge as a starting-point for our analysis. However, in order to do justice to the teachers' practical knowledge (...), we think that the use of insights from other research is only legitimate when this is done on condition that the insights can be reformulated or revised in order to make them fit the data. (2002: 162, emphasis added)

Patton (2002), referring to these as sensitizing concepts, provides a rather interesting example of a study in which concepts normally associated with one group were applied in an investigation of a contrasting group: the concept of victimization, often applied to those harassed by police officers, was used instead to study police officers, giving a new perspective on their experience.

In vivo, or indigenous codes

In direct contrast to *a priori* codes, *in vivo* codes are derived directly from the data (Strauss, 1987), capturing an actual expression of a participant as the title for a code. Such codes contrast with sociologically or theoretically constructed codes, and reflect an 'emic' approach to analysis.

When a member of a corporation which had undergone a radical change program talked about the 'hard labour' of working through that change process, this expression was appreciated for its valuable imagery – of what goes on in a birthing process, of the slogging work of someone held in a prison – and hence became an in vivo code available for use in coding other text.

Patton (2002), in a similar vein to Strauss, uses the idea of indigenous typologies as a way of seeing how participants break up the complexity of reality. He suggests looking for local terms, especially those that may sound unfamiliar or are used in unfamiliar ways. Using the language of the participants to label the typological concepts, he then explores the dimensions of those concepts with the participants to identify the attributes or characteristics that distinguish one from another. For example, with teachers talking about dropouts, he might ask what separates a 'chronic' from a 'borderline'. Patton suggests that use of indigenous terms and typologies also makes feedback more comprehensible to participants.

> Codes created directly from the words of the participants (*in vivo* codes) can sometimes have the unfortunate problem of not being useful in exactly that form for expressing what is learned from other participants. Again, feel free to change the term to a more general construct as the project develops, but keep a record (in the description, or a memo) about how the code arose in the first place.

Repetitions and regularities

Strauss observed to Lyn Richards that possibly missing coding an instance of something that someone said is not a cause for anxiety: anything that is important will come up again.[4]

People repeat ideas that are of significance for them (Ryan & Bernard, 2003). Repetitions therefore suggest useful concepts to use as a basis for nodes.

> Both Frank and Elizabeth, despite their very different pathways into research, repeatedly emphasised their need for *disciplined time management*:
>
> Frank: you've got to be disciplined in how you manage time…you've got to set your objectives and set your timetable and build in your objectives
>
> Elizabeth: I have certain days of the week just for research days and it is amazing just how many things that just don't have to be done … it gave me really two days free to work on research and writing and you have to be ruthless about it … so it meant that I really had to be quite ruthless.

Regularities in the text may suggest patterns of association that are worth catching as well. If there is an overarching concept involved, name it as a separate node; if the association involves a cluster of nodes, make a note of what these are for later verification (when you have furthered your NVivo skills, you might use a ***set*** or a ***relationship*** to link them).

Enjoyment of research was often linked with being intellectually stimulated by it. This warranted a note in my journal, and later I used coding queries to verify the association.

Ask questions

Use questions of the text to generate codes – who, what, when, why, how, how much, what for, what if, or with what results? Apply the Strauss (1987) coding paradigm to generate ideas that relate structure to process (without necessarily assuming cause and effect): What actions or interactions are occurring? What strategies are being applied? Under what conditions? With what consequences? Each aspect then warrants a separate code. Asking these kinds of questions will help also to ensure thoroughness of coding and to develop relational statements (do the actions or strategies change under different conditions and if so, what are the implications?) and so will hugely benefit (and simplify the process of) development of a theoretical model.

Elizabeth talked about the context and impact of discovering that research could have social justice benefits: "... that gave me a whole new focus on research because here people were doing research that really affected people's lives ..." Asking questions about her report of her experience reveals a range of possible codes:

- This represented a *change in perspective* from a *discovery model* of research, which had been acquired from her parents, to a *social benefit* model;
- It arose in the context of her *work experience*;
- While clearly her interest in research was *rekindled* (I initially used raised, but then noted that she went on to say that she guessed "it had always been there") ...
- ... there was no immediate effect (she went off and became a depressed housewife), but it did lay a *foundation* for future job-seeking as a researcher (although a very different kind of foundation from that talked about by Frank); and
- It was a *turning point* (or *critical experience* or *milestone*) in her path to becoming a researcher.

Additionally (or alternatively), one might ask whether terms (discourse) used in the text reflect a particular construction of the topic, or of society. What has led to these constructions? What are the implications of seeing the world in this way?

> Amongst some more experienced researchers one can detect a discourse of *performance* (being successful, competing, building a reputation), and alternatively, of *play* (following curious leads, puzzling, playing with new ways of doing things) in the way they talk about their engagement in research activity. Coding these varying constructions of 'doing research' will facilitate assessment of what they mean for their work and career.

Compare and contrast

Compare one segment of text with another. Think about the ways in which they are both similar and different. This will help you go far beyond simply deciding which chunks might go together: it will help you discern the dimensions within, or perhaps to discern previously unobserved variables running through the text.

In a study of parental and peer attachment among Malaysian adolescents, Noriah Mohd Ishak, in the Faculty of Education at the University of Kebangsaan, Malaysia, found that boys and girls spoke in contrasting ways about their parents' role in their relationship choices, for example:

> Boy: I trust my parent with their choice, I might have a girlfriend here in America, but to whom I will finally get married, will depend on my parents choice. I think theirs is always the right choice.
>
> Girl: I choose whom I will marry. My parents might have their own choice, but how can I trust their choice, because they live in a different generation than I do!

In relation to young people's acceptance of parental leadership, then, trust is a significant dimension. Other possibly relevant concepts arising from these comparative passages are parental adaptability, parental authority, cultural expectations, cultural change. Also of interest in the boy's comment about having an American girlfriend is what it says about attitudes to women and commitment in relationships.

Don't limit your comparisons to actual examples. Grounded theorists suggest you bounce off from the text into hypothetical comparisons – perhaps quite extreme comparisons – in order to explore more deeply the structure and significance of the words used and to identify their properties and dimensions (Strauss, 1987; Strauss & Corbin, 1998).

> As his career was unfolding, Frank's total focus was on research: "When you go out in the pub you are talking research, when you go to bed at night you are thinking it." How does this experience reflect or contrast to that of an addict? This prompted me to talk with a colleague who was studying youth gambling behaviour. Researchers and gamblers alike follow a passion, often (but not essentially) to the detriment of their health and/or family relationships; both provide financially insecure career paths which can benefit from careful strategic planning but which are also subject to 'luck' or whim; both can be characterized by emotional highs and lows; gamblers experience problems with self-regulation. We argued over whether addiction was necessarily 'bad'. As the project developed further, I saw addiction as one expression of a larger category of 'obsession', which I came to define as 'driven passion' to capture the dimensions of emotional engagement and potentially blind (unregulated) commitment of the obsessed researcher.

Comparative techniques help move attention from factual description to increased sensitivity to the dimensions of the concepts being derived from the data (such as the drivenness of obsession; and its emotional volatility as fortunes in competition wax and wane), as well as to overcome analytic blocks. One consequence of doing so is that you may create nodes simply to hold ideas, for example, as sociological, 'hypothetical' or 'logically derived' nodes (Strauss, 1987). If these nodes never acquire any data, then that is of interest too.

Record narrative structure and mechanisms

Interpretive analysis takes account of more than the content of what was said. How things are said, and the way in which the text was structured by the interviewee – the discourse and narrative features of the text – are also revealing. Telling a narrative brings order and meaning to events. Conversations have structures. Linguistic expression reflects wider discourses in society. In emphasizing the social character of communication, Drew (2003: 141) noted that "language is employed in the service of doing things in the social world" and in consequence it reveals how participants in the conversation co-construct meanings of social interaction.

Particular features that you might note (and annotate or code) include, but are not limited to:

- transitions and turning points in the narrative, signifying a change of theme or a subject to be avoided (Lieblich *et al.*, 1998; Ryan & Bernard, 2003);
- inconsistencies, endings, omissions, repetitions and silences (Poirier & Ayres, 1997; Silverman, 2000);
- denotations in time, and tenses in verbs, indicating identification with the events described and/or attempts to distance an event or bring it closer (Lieblich *et al.*, 1998);

- the use of metaphors and analogies (Coffey & Atkinson, 1996; Patton, 2002; Willig, 2003);
- the repetitive use of a word or phrase (Ryan & Bernard, 2000);
- the structural aspects of turn taking and other 'rules' in naturally occurring conversation (Drew, 2003; Peräkylä, 2004; Silverman, 2000) including, for example, the sequence leading to the initiation of an action (such as offering an invitation or advice, or making a request) and its relation to the design of that action (e.g., whether the request is direct or hedged; whether the invitation is informally or formally delivered; whether it offers a way out);
- the broader discursive construction or framework within which the discourse is set, for example, biomedical, romantic, or gender (Willig, 2003);
- narrative (story) components within a longer non-narrative text (Riessman, 1993);
- the sequenced, structural elements of a narrative (Elliott, 2006; Riessman, 1993); and
- use of particular articles or pronouns pointing to particularized or generalized referents, for example, the staff, my staff or our staff; indicating level and type of ownership or involvement (Lieblich *et al.*, 1998; Morse & Mitcham, 1998).

Just as using cases as your unit of analysis does not necessarily imply you are doing a case study, observing narrative features does not necessarily mean you are doing a narrative or discourse analysis. These are elements in the text which can potentially inform your analysis, regardless of your methodological approach, and so my suggestion is that you create nodes specifically to identify interesting narrative features of the text (these may become a 'stand-alone' tree of nodes, *cf.* Chapter 5). Of course, those who are undertaking conversation, discourse or narrative analysis will be analyzing features such as these (and other similar ones) in much more detail (*cf.* Chapter 8).

CONCEPTS, CATEGORIES AND THEMES

Concept, category and theme are words that are used often and inconsistently in social science methods literature.

Strauss and Corbin describe a *concept* as "a labeled phenomenon...an abstract representation of an event, object, or action/interaction that a researcher identifies as being significant in the data". Concepts with shared properties are then classified into *categories*. Categories are also concepts, but they are ones which "stand for phenomena" (i.e., more than one object). The same concepts may be classified and labelled in different ways, depending on which properties of the concepts are being considered. The category then has more explanatory power than the concept (Strauss & Corbin, 1998: 103, 114). In contrast, Howard Becker, another symbolic interactionist, sees concepts as "generalized statements about whole classes of phenomena ... that apply to people and organizations everywhere" (1998: 109). Concepts arise from "continuous dialogue with empirical data" as a way of summarizing that data through systematically relating criteria around a central issue.

I asked Barbara Bowers (University of Wisconsin Madison) how Strauss had used the terms concept and category in his teaching when she was a doctoral student at the University of California, San Francisco. Her response (personal communication, 14/09/2005):

> First, Strauss was not consistent in how he used these terms. The best answer I can give you is that both terms were used to refer to multiple levels at different times. I have even found things he has written that had property as a more abstract term than category. In the Strauss and Corbin book he was forced for the first time to come to terms with how he was using these different terms. He was more consistent in this book, but not necessarily consistent with other things he had written. He actually got angry when students asked these questions. He didn't see it as particularly relevant or important. 'Call it what you want.' It was the process that he was concerned about.

Because I link *concept* with the act of conceptualizing and have tended to think of *category* as a descriptive label, I think of each as having a place as different but necessary building blocks for analysis, but with concept as being of a higher order (of abstraction) than category. I use the terms together quite frequently, and use them more or less interchangeably when talking about the products of coding.

Ryan and Bernard (2003: 87) define *theme* in much the same way as Corbin has described a category and Becker a concept:

> ... themes are abstract (and often fuzzy) constructs that link not only expressions found in texts but also expressions found in images, sounds and objects. You know you have found a theme when you can answer the question, "What is this expression an example of?"

In contrast, Richards and Morse (2007) see a *theme* as something which is more pervasive than a topic or category, something that runs right through the data and the creation of which involves abstract thinking and copious reflective memoing.

I have a problem with the popular use of the word *theme*, born of reading too many research proposals which have summed up analysis of interview or focus group data (actually, often described it in full) as 'themes will be identified in the data'. The implication that a theme is an idea which runs through the data is not a problem, but if this way of describing analysis means that nothing further is asked of the data beyond identification and description of these pervasive ideas, it is very much a problem. As a consequence, I use the term infrequently, and become suspicious whenever I encounter it.

STORING CODING IN NODES

As noted earlier, coding in NVivo is stored in ***nodes***. In information systems the term 'node' is used to indicate either a terminal point or a connection in a branching network. Sociologists might be familiar with the idea of nodes in a social or a semantic network. Horticulturalists know the node as the point at which branching might occur in a plant. Similarly, in a fully developed NVivo coding system, nodes become points at which concepts potentially branch out into a network of sub-concepts or dimensions.

In NVivo, you make a node for each topic or concept to be stored, much like designating a hanging file for each topic. What NVivo keeps there, however, are not actual segments of data, but *references* to the exact location of the text that you have coded, from the source document. Using those text references, the software is able to locate and retrieve all the coded passages from the document records. The passages themselves are never copied, cut or physically moved into the nodes. Unlike cut-up photocopies on the lounge-room floor or in hanging files:

- the source always remains intact;
- information about the source and location of a quote is always preserved;
- it is always possible to view the coded passage in its original context;
- changes to the document are immediately reflected in the text viewed through nodes; and
- passages can be coded at multiple nodes, with queries able to find passages coded by co-occurring nodes.

At first, you will probably use ***free nodes*** to store your coding. Free nodes do not presume any relationships or connections – they serve simply as 'dropping-off' points for data about ideas you want to hang on to. Later these are likely to be organized and moved into ***trees*** – hierarchical, branching structures in which ***parent nodes*** serve as connecting points for subcategories or types of concepts (the topic of tree nodes will be covered fully in the next chapter).

In most cases, nodes in NVivo will store coding about topics or concepts or themes, but sometimes nodes are used for organizational purposes (such as to store case information). This is because using a node is the best way to bring together different segments of text, whether they are from within a single source or from different sources. Additionally, some nodes store no information at all;

for example, parent nodes in trees often have no coding, or you may create *a priori* or 'logically derived' nodes for which you do not find data. The number of sources and passages (references) coded by each node is shown in the List View display of the nodes, along with their creation and modification dates.

MAKING NODES AND CODING

There are multiple ways of making nodes, either as you are working through the text or when you are just thinking about the categories you might need.

- To begin making nodes and coding your data, select the **Nodes** tab in the Navigation View for your project, *while keeping a document or memo open in the Detail View*. Select **Free Nodes**.

To create a new node and code text at the same time, choose from one of the following methods:

- Select text, click in the code slot in the Coding toolbar (or double-click to select and overtype what is already there), type a name and click the Code icon
- Select text, and from the RMB, choose **Code > Code Selection at New Node** to open a new node dialogue. Type a name and press Enter. Add a description if you wish.
- Select text and press **Ctrl+F3** on your keyboard, to open a new node dialogue.

If you select text and then forget to click in the coding slot before typing, you will overtype your selected text (use Undo to recover it).

If Tree Nodes is showing in the Navigation View, you may find you are making new top-level tree nodes rather than free nodes (although you might want to do this deliberately).

Watch that you don't accidentally code the whole source.

To prevent text from being inadvertently moved (and also to see that your codes are being attached to the text), turn on the coding density bar (**View > Coding Stripes > Show Coding Density Only**).

We've all had the annoying experience of losing control of the text selection when the mouse is moved below the base of an open window. To prevent this happening, click where you want the selection to start, use the scroll bar or your roll button to move further down in the text, then hold down your Shift key and click again where you want the selection to finish. All text between the two clicks will be selected.

Adding further coding

Detailed coding will be quite slow at first. Not only are you orienting your thinking to the issues raised by the data – which is likely to mean spending time in reflection and in adding to memos and annotations – but you have also the practical task of creating new nodes as you work, and making decisions about those. So, while new ideas are exciting, you may also be feeling anxious about how much time this is all taking. The number of new nodes you are creating will decrease markedly after the first two or three documents, and once you create a node, you can access it easily for further coding. As the text referenced by your nodes builds up, your coding will start also to develop a clearer structure, you will have gained more confidence in working with your data, and you will find your pace increasing.[5]

For the first document (or two), you are most likely to add to the list of free nodes. Trees will come later, as you gain a sense of the emerging structure of your data (*cf.* Chapter 5). The effort that you put into creating free nodes is well spent: free nodes allow you to capture ideas (and text) without forcing structure too early, and all the coding you are doing will stay with the nodes when you are ready to rearrange them into a structured tree system.

There are several ways to continue adding coding to documents. One of the 'problems' with NVivo is that there is more than one way of achieving most tasks, and it takes a deal of experience to know which most suits your purpose at a particular time. For now, you need to find a method of coding that you are comfortable with, because you're going to be doing quite a lot of it! As your project grows, you may find another way to work, one that better suits your needs at that point.

ADDING FURTHER CODING

Before you start:

- Rearrange your screen for drag-and-drop coding: Click on the View toolbar, or choose **View > Detail View > Right**, then move the pane divider toward the left.
- Turn on the coding density bar: **View > Coding Stripes > Show Coding Density Only**. This renders the document read-only, which means you cannot accidentally edit it while dragging text to a node. It has the added advantage that if you hover over it you will see what coding you have added to the adjacent text.

The easiest methods for now are likely to be:

- With nodes showing in the List View, and the text you are coding in the Detail View, drag selected text to a node.

▶ Select a recently used node from the drop-down list on the Coding Bar and click.

Alternatively, if your focus is on, say, writing up field notes or writing memos, or you want to code at multiple nodes at the same time, you may find it easier to:

▶ Highlight a passage (in Detail View), then select a node (or nodes) for coding using the coding toolbar (recently used nodes OR Select Nodes), OR **Ctrl+F2**, OR **RMB > Code > Code Selection at Existing Nodes** (Figure 4.2). These options are available regardless of what is showing in the List View.

If you have already used the node selection dialogue in this working session, and you are now choosing different nodes, click Clear before making your new selection/s.

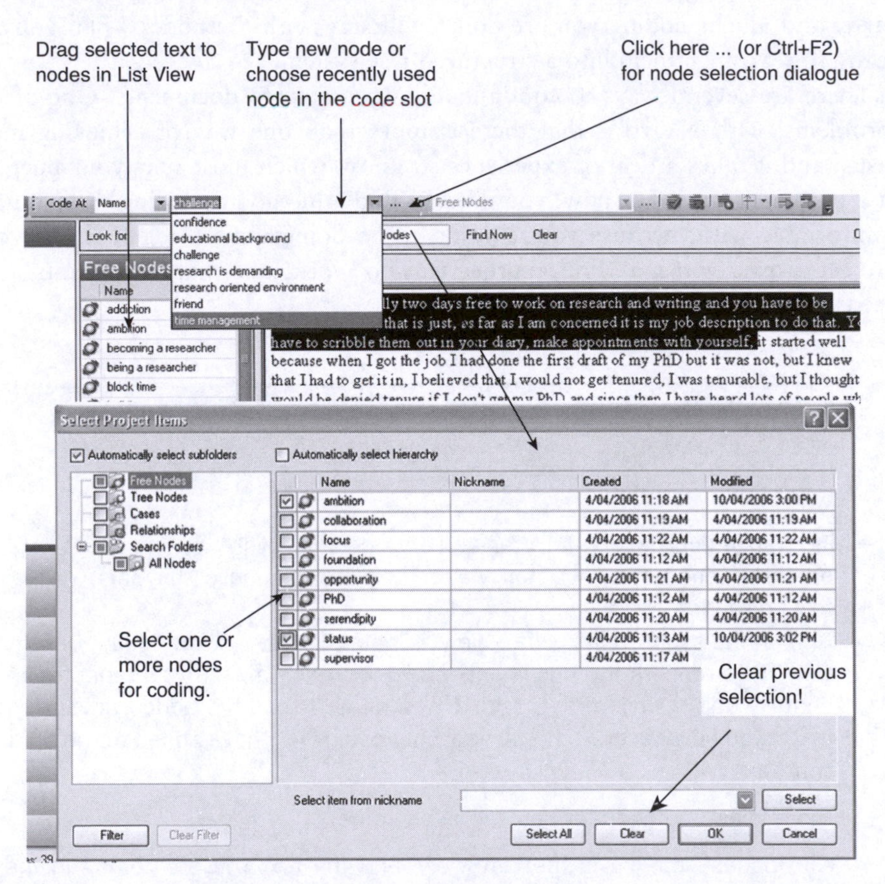

Figure 4.2 Alternative ways of adding further coding

- Nodes in List View can be reviewed while you are thinking about which to use for coding: double-click on the node you're thinking about, review its text, then close the Detail View by clicking its close box, to return to the source text. Alternatively (or as well), open its **Properties** (Ctrl+Shift+P) and check (or add to) its description.
- Coding can be 'undone' on a selected passage either by choosing to **Uncode** (at whatever node is selected), or by immediately clicking **Edit > Undo**.
- Nodes can be deleted (**RMB > Delete** or select and click Delete on your keyboard), thus deleting all coding done at them.
- You can code at multiple nodes at once by using opening the selection dialogue and choosing as many nodes as are required (make sure you Clear before making a selection for a new passage).

When it's time to open a second document, notice that the nodes remain in place, ready to use with the new document. You will select from the same (expanding) list of nodes to code all your documents – no need to create a new version of existing nodes each time you start on a new document. There is also no need to create a new version of the same node for documents that come from a different subgroup of your sample – those differences will be handled by ***attributes***, and it makes really good sense to have *everything* you know about a particular concept or category stored in the *same* node, regardless of circumstances. The query tool will allow you to see how that concept looks for different subgroups.

Checking as you code

Often there is a need to check just what coding you have already added to a passage:

- You're still getting used to the software and want some reassurance that your coding is really happening;
- You've been interrupted in your work and want to check where you are up to; or
- You want to check if you have picked up all important aspects of this segment of text.

You can achieve these various goals in several ways.

VIEWING CODING

Using coding stripes

- Use the coding density bar to see what nodes code the adjacent text, and where coding was stopped.
- Right-click and choose **Show Stripe** to view the occurrence of a particular node across the whole document.
- View coding stripes for the whole document or for a selected passage: Ask for the **Nodes Most Coding Item** (or **Selection**). Up to 7 stripes will be shown. You can change the number shown using the Coding Stripes menu (**Number of Stripes...**).
- With coding stripes displayed, click on a stripe and access the RMB for further (sometimes useful!) options.

Reviewing text coded at a node

- Check the text coded at a node by double-clicking the node in List View. Passages coded will be displayed in Detail View. Options for detailed review of nodes will be fully explored in Chapter 7, but in the meantime, you might make use of the options for viewing the context of the node on the RMB menu (*cf.* Figure 1.2 in Chapter 1). These will allow you to review text in its original source (e.g., paragraph, or whole document).

Automating routine coding

A significant amount of what any researcher does is routine work – the 'dogsbody' tasks which are boring but essential to the overall project. Licking stamps and sealing envelopes is one I remember from my (pre-web) days of running surveys, along with the tedious but essential task of entering numeric codes onto the computer. Fortunately, for researchers and the progress of research, technology has come to our aid and machines have replaced some of the drudgery.

Even in qualitative analysis, the researcher can automate routine coding-related tasks, giving more time to concentrate on less mechanical, more interpretive work. In NVivo, using the ***auto code*** tool is the primary way to achieve this. For example, use auto coding to:

- Code responses to standardized questions, such as are generated by self-completed questionnaires, for the question to which they were a response;
- Code passages for topics identified by headings in the text;
- Delineate the components of regularly structured documents, such as submissions or annual reports;
- Delineate the contributions of each speaker in a focus group, meeting or multi-person interview, in order to code each speaker's text to a case node;

- Identify cases, where everybody's responses to one or more questions are in one document.

The resulting nodes will give immediate access to, say, all the answers to Question 3, all submissions about Eligibility Criteria, or everything that Jim said in the focus group. Critically, being able to do this depends on your having used headings to identify the relevant sections in your sources (*cf.* Chapter 3).

Auto Code by Heading Level is the principal and most useful method of automating coding.[6] When you auto code documents using headings, text under a heading (and before the next heading) is coded at a node with a name based on that heading.

AUTO CODING SOURCES

▶ In List View for your Documents (or a sub-folder), select the source or sources you wish to auto code (these should all be of the same general type). If you are auto coding surveys or questionnaires, then select the whole set at once. If you are auto coding focus group transcripts, it will be safer to do them one at a time, especially if your method of identifying different people is repeated in different groups.

▶ Choose to **Code > Auto Code**, or click on in the Coding toolbar. Choose the level of heading identifying the text you wish to code, and where you want the resulting nodes to be located. Note that, for both Trees and Cases, you will need to create a parent node to 'foster' the new nodes.

✓ For survey questions where each case is in a separate document, you are most likely to want to code for **All** headings at the same time. This will produce a node structure which replicates the structure of the survey (Figure 4.3).

✓ For focus groups which have headings for both topics and speakers, you will want to code for particular levels of heading in separate passes (see *Researchers* example, below). Similarly, for survey documents created using a Merge file and comprising multiple cases and repeated questions, auto code in two passes, one for cases, and one for questions.

✓ If auto coding produces a node which contains headings only (e.g., for *Thinking about vaccines* in the immunization example), and you wish to see the text of all the subsections, then in that node: Edit > Select All; and then RMB > Spread Coding > Heading level. This will permanently spread the coding for that node to include all subsections.

- If the nodes are out of order, select the parent node and then click on Sort By Custom.
- If you mess up, simply Undo, or delete the nodes you have created and start again!

Check **auto code** in **Help** for additional information and examples.

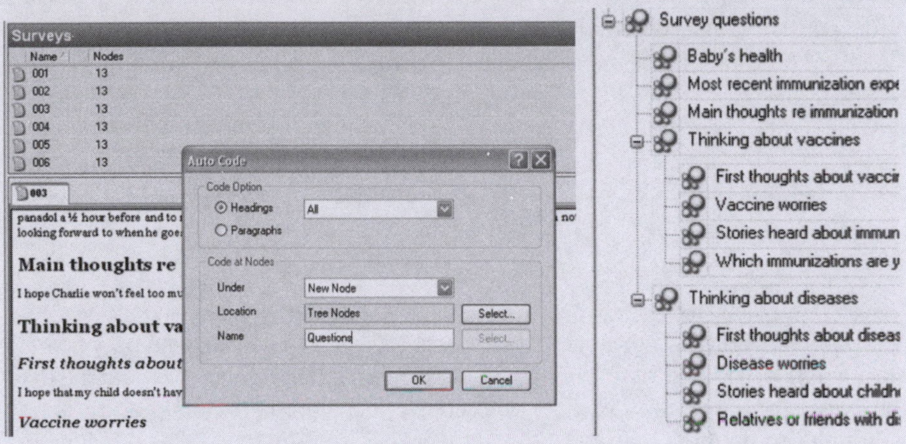

Figure 4.3 Auto coding questionnaire responses

Where the text for questions or topics in a structured survey has been auto coded and the data from demographic or other categorized questions have been imported as attributes, then responses to particular questions can be compared immediately for different subgroups, using the matrix coding query tool (*cf.* Chapter 6 for use of attributes and matrix queries).

In focus groups or multi-person interviews where each participant can be individually identified, autocoding for speakers is the most efficient way of creating case nodes for each participant. These will be necessary for storing attribute information relating to each of those cases. Also, in an auto coded focus group, you will be able to quickly check if it is always David who talks about self-indulgence, or whether several people referred to research as self-indulgence.

Several of the focus groups have Level 1 headings for topics and Level 2 headings for speakers. The most appropriate way to auto code them is in two runs. First, I coded each for Level 1 headings at a tree node for Topics; then for Level 2 headings, at a case node for Group Participants. This results in all the text being coded twice, once for topic and once for which person is speaking. A tree of nodes covering each topic is created; and a node for each member of the group is created under Cases\Group Participants.

If, instead, I had auto coded a group document once at All Headings, the topic nodes would contain just the topic headings, with no other text, and the speaker nodes would have been repeated under each topic. Neither of these outcomes would have been helpful.

Autocoding is less useful for single-person interviews, although one of the reasons I typically use Heading 2 for speaker names is so that Heading 1 is still available should I wish to put topic headings into my documents (see, for example, Elizabeth's document in the Researchers project, which contains a number of excerpts from a much longer interview). In most interviews, even those where a semi-structured interview guide is used, the discussion rarely sticks closely to set topics, or it covers several topics at once, and so it often isn't appropriate (or practical) to use headings in the interview transcript to identify topics. One of the issues in trying to insert headings into a conversational document is that once you start doing it, you are more-or-less bound to continue, otherwise all remaining text is contextualized by the last heading used.

Other coding tasks can also be semi-automated in another way, by using the capacity of the program for searching text and saving the found passages as a node. ***Text search***, as a strategy for locating, viewing and coding text, will be covered in Chapter 7. As a tool for automating coding, text search can:

- Code passages based on repetitive features, such as a common term appearing at the start of a paragraph or section;
- Code topics or people or groups identified by a keyword, for example, in a tourism project, identifying passages (paragraphs) about regularly mentioned locations or environmental features, or, in a project about care of infants, identifying those paragraphs in which the major activities of feeding, bathing, or sleeping were discussed.

✔ Most features that support auto coding, particularly use of heading styles, are best planned for at the stage when you are preparing your documents. Once the document has been imported, it is much more time consuming to incorporate the necessary features. If you are in a situation where you would like to apply heading styles to question titles or speaker names after the document is imported, however, you can potentially use the **Edit > Replace** function to modify the text and apply styles as you do.

✔ If your project doesn't have particular features which readily support auto coding (for example, there aren't regularly occurring topics or questions which can be identified by headings), don't force the issue by attempting to apply auto coding strategies. It will be more efficient and effective to code the text interactively.

Coding offline

This is *not* my recommended way to go about coding. There are just so many advantages in coding directly on screen in terms of thoroughness, efficiency, and ability to simultaneously record or explore ideas. For those who want to make use of the time opportunities provided by travelling in public transport, however, or who for some other reason can't work on screen, doing some or all coding initially on paper may be the way to go. If you do code offline, then the coding can be added to your documents later either by working interactively with the text (i.e., as you would normally), or by using the paragraph coder.

There are three significant issues with working on paper:

- It is difficult to clearly mark on paper which components of the text belong to which codes, especially in rich text requiring multiple, overlapping codes (unless you are selecting whole paragraphs only).
- Most obviously, the coding will still need to be transferred to the computer, so although the thinking about codes may have been achieved while offline, there remains a rather boring, mechanical task of transferring that thinking onto the computer. In some cases, this can take as long as it would have done to code directly on computer in the first place.
- If you are using the paragraph coder to enter the coding in the computer, codes can be applied to *whole* paragraphs only.

✓ If you are coding offline because you can't access your computer at the time, use the opportunity primarily to 'rough out' some coding ideas for the text, supplemented with marginal notes. Then when you come to transferring it, take a second look and reflect a little more on the detail, as you add the coding in the normal way, and transfer your notes to memos.

✓ If you plan to have someone else (an assistant?) transfer your work to the computer by selecting text and coding, then apply codes to whole meaning units (or multiples of) so that it is clear just what text is to be coded at which nodes. 'Meaning units' would normally be indicated by paragraph breaks, but slashes inserted in the text might also be used to clarify where the breaks or changes in coding are intended.

PREPARING FOR AND USING THE PARAGRAPH CODER

In order to prepare for using the paragraph coder, you will need to create a copy of the document from NVivo, printed so that it has paragraph numbers on it (**RMB > Print > Print Document**). You will also need a printed list of your nodes (*cf.* Chapter 7 for advice on the best format if your list includes tree nodes).

- You are likely to code by writing node names in the margin on the printed document. This is fine for transferring coding on-screen if you plan to use the regular method of selecting text and coding to enter it. For transfer into NVivo using the paragraph coder, however, it is more efficient to work with a printed node list (one copy for each document to be coded – make sure you write the document name at the top). Find the node in the list, and write in the paragraph number or range after it, to indicate where that node applies in the text (Figure 4.4). Add new nodes to the list as needed, and add further paragraph numbers, as appropriate, to the line beside each node name. This will speed up data entry enormously.
- Select the document you want to code (List View).
- Open the paragraph coder by going to **Code > Paragraph Code** or by clicking on the paragraph code icon in the coding toolbar. In the paragraph code dialogue, select a node (or nodes) you wish to code at, and type in the paragraph numbers or ranges of text for that node (Figure 4.4). (Tick them off on your list as you do!)

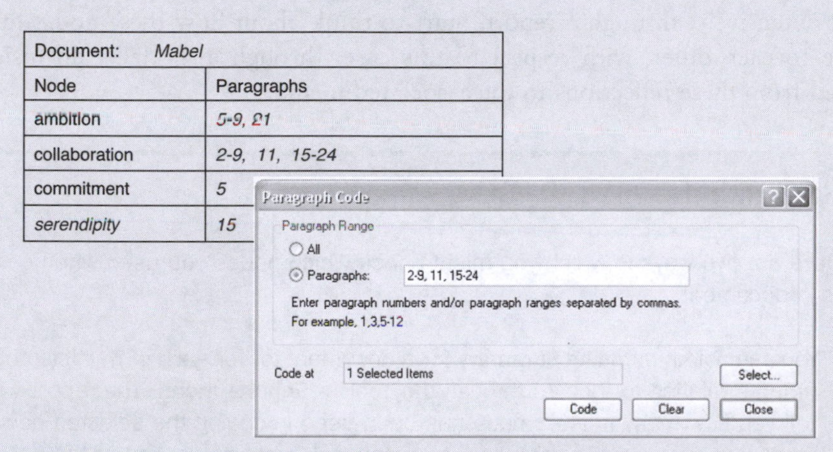

Document: *Mabel*	
Node	Paragraphs
ambition	*5-9, 21*
collaboration	*2-9, 11, 15-24*
commitment	*5*
serendipity	*15*

Figure 4.4 Recording sheet and paragraph coder

What's happening as I code?

As you code, NVivo is adding references to the source text at the nodes you are using, information which is stored in the project database. In the process, you are connecting your emerging ideas about the concepts in your study to the data about those concepts. Not only will all the text (or other content) you have been coding be accessible from the nodes (with each passage carefully identified by NVivo as to where it came from), but any annotations, memos or other files linked to it will be accessible along with the text. Reviewing the text stored at a node, therefore, will allow you to review, as well, *all* the associated ideas linked to that text.

NVivo is keeping a tally of the coding you are doing, so it is possible as part of the reviewing process to obtain a report on the number of sources, cases, paragraphs, passages and words coded at each node (**Tools > Reports > Node Summary**).

NVivo stores the same text reference once only at a node, so if you code the same passage twice at the same node, the text at the node is *not* doubled. Similarly, if you ***merge*** the contents of two or more nodes (or use the query tool to combine nodes) and the same passage was coded at more than one, it will appear once only in the merged node. If you edit coded text in the source document, those edits will also be reflected in the text retrieved through the node.

REFLECTING ON THE CASE

You have been working intensively with the content of a document. Now it is time to gain a little distance, to try to put what you have learned from this source or about this case into perspective.

Before moving on, take a little time to reflect on what you have learned from the source or case you have just been examining. Check which nodes were used most extensively, through a report. Start to think about how these nodes might relate to each other, with respect to this case, through a model. Add insights gained from these reflections to the associated memo.

WHICH NODES DID YOU USE?

There are two ways in which you might check which nodes you used when coding a document.

- You can obtain a **Coding Summary** for a document (or for each of multiple documents relating to a case) through the **Tools > Reports** menu. The report will tell you how many times a node has been used in coding the selected document (i.e., number of references, or passages), what proportion of the document is coded at each of the selected nodes (or all nodes) in percentage terms, and will optionally include each passage coded by each node.[7]

✔ Watch what you ask for! If you ask for text, then do not include case nodes in your request. Be aware that each passage will be repeated for each node you have used in coding it.

- Identify the nodes coding a document (or a case) using the Find toolbar at the top of the List View. Go to **Options > Grouped Find**, ask for **Items Coding**, and select either the document or case as the **Scope** item and Free and Tree Nodes as the **Range** items (Figure 4.5).

✓ If you want to use these items in a case-based model, then **Edit > Select All**, and **RMB > Create As > Create as Set**.

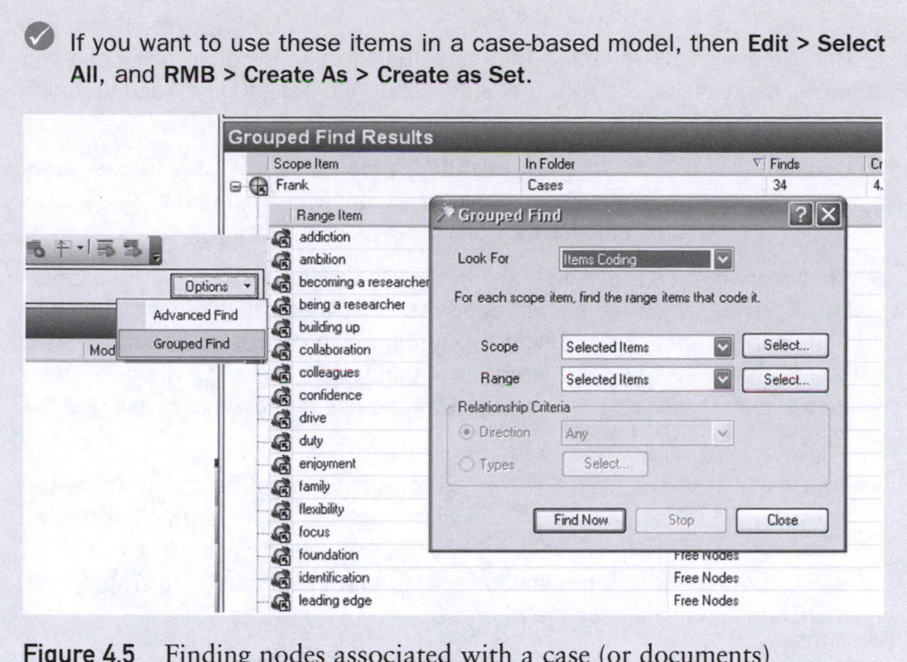

Figure 4.5 Finding nodes associated with a case (or documents)

Modeling a case

Try visualizing how you see this case, using a model. Models are where you can best gain a sense of how the various nodes you have developed might fit together to tell a story – the story of your research question from the perspective of this participant or case.

MODELING A CASE

For general assistance in using the modeler, go to **Help** and, using the **Index**, select **models** and follow the links. The following steps outline how you might approach the task of modeling a case:

- Select **Models** in Navigation View. Create and name a **New Model** (RMB in List View). Undock the model window and expand to get a full screen view to work in (toggle **Window > Docked**). Turn off the Groups panel (**View > Model Groups**);
- From your RMB, select **Add Project Items.** Choose to add the **document** you have been coding, and **Add Associated Items > Items coding.** Click on the document's shape and delete it to leave behind all the nodes coding that document;

OR
Use **Grouped Find** to locate and show all the nodes coding the case you have been working on, save these as a Set, and bring the set (and its members) into the model.

- Delete any nodes that are not useful for presenting this case. Move items around to best illustrate the process or experience described by this participant (or fitting this situation – whatever the case might represent).
- Show ***connections*** between items in the model (remember that the connections you are showing in *this* model are just those that reflect what was happening in *this* case): Select two items using **Ctrl+Click**, hover over one of the items and select **RMB > New Connector**. If you create a one-way arrow that is pointing the wrong way, select it, hover and use **RMB > Reverse Direction** to fix it.
- Adjust font type and size, fill and line colours for a selected item using the **Format** menu. You can adjust the size, font, or colour of multiple model items at the same time: Select those you want to change (drag to select; use **Ctrl+A** for all; or **Ctrl+Click** for some), and then adjust one. The adjustment will apply to all selected items.
- Enhance your model's power to explain and communicate by applying ***styles*** to the shapes within it (these work on the same principles as heading styles). These can be set up for this project through **File > Project Properties > Model Styles.**

✓ You can create a static view of a model if you want to preserve a copy to reflect a particular case or stage in your thinking, before moving on to develop it.

Two early case-based models are shown in Figure 4.6. The model of motivation to be a researcher for Frank has been drawn from a grounded theory perspective, while that for Josie was designed to help identify essences of her experience as a research student, from a phenomenological perspective.

Adding to the document (or case) memo

Record the thoughts prompted by your review of coding for the document, and those generated while building a model to represent what was happening for this case.

Sum up what you have learned from this case. In this final memo for the case, try to move from a descriptive to a conceptual level. Focus on ideas generated in relation to your research questions, rather than simply summarizing around specific events or people.

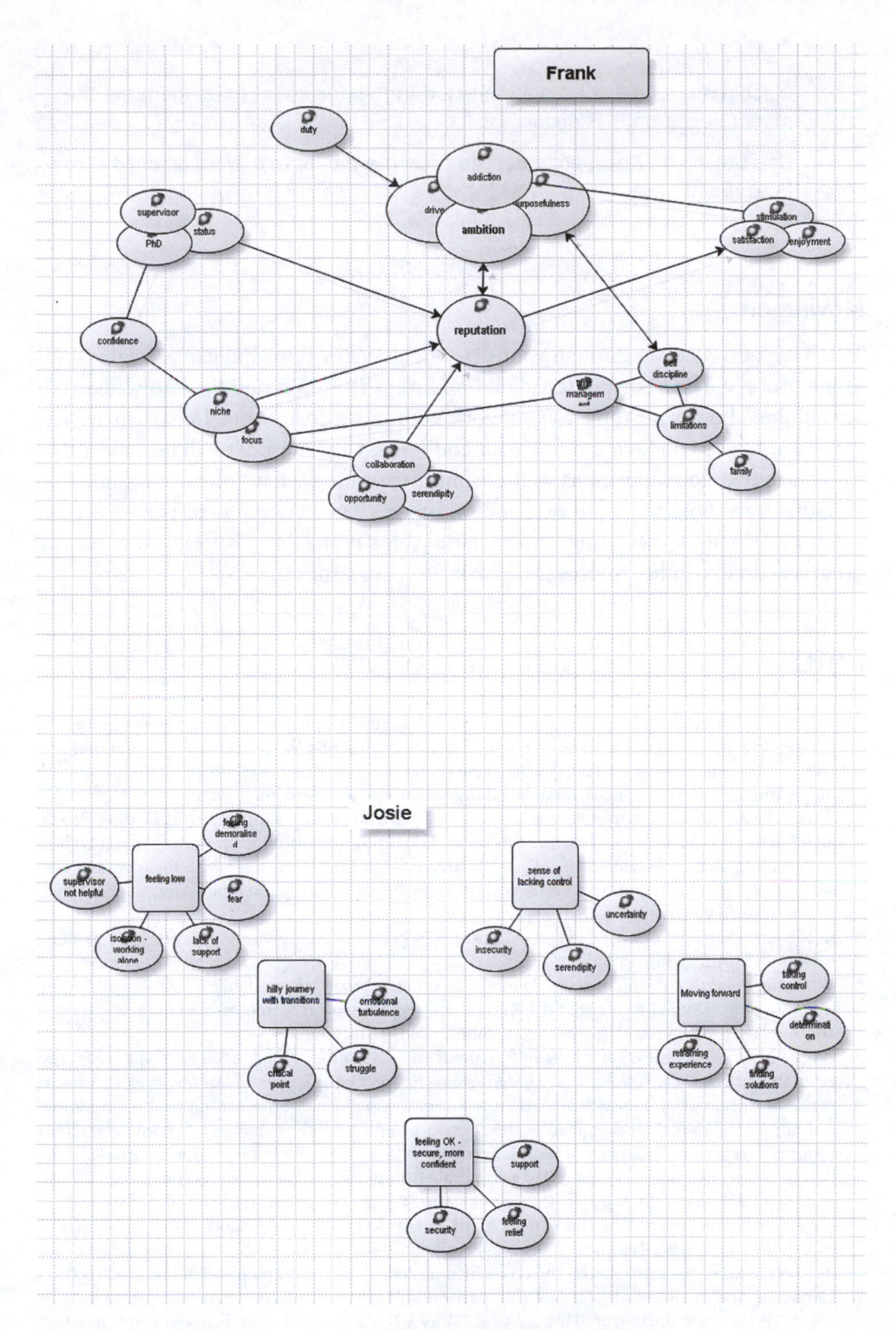

Figure 4.6 Two types of case analysis using the modeler

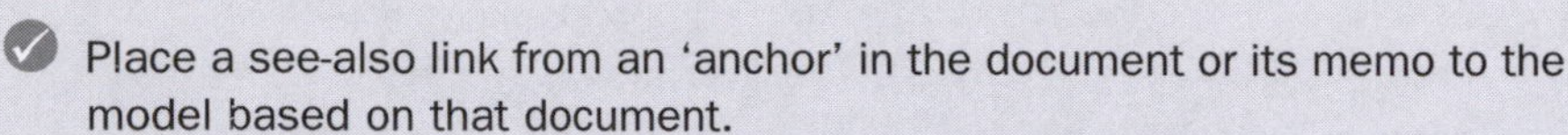
Place a see-also link from an 'anchor' in the document or its memo to the model based on that document.

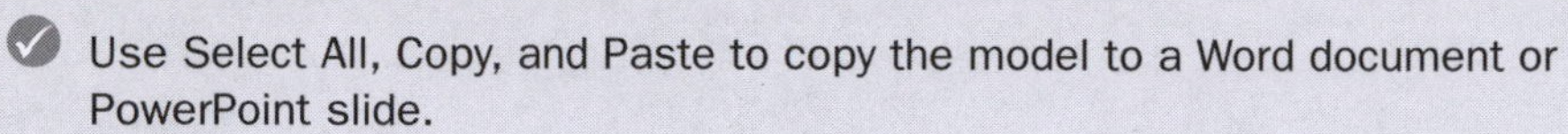
Use Select All, Copy, and Paste to copy the model to a Word document or PowerPoint slide.

Moving on

You have now worked through your first document or two, and gained many skills in the process. The next step is to build on both these skills and the knowledge you are gaining about your topic as you work with more documents. In particular, you will learn how to build and refine your coding system, so that it will become a tool for conceptual development as well as simply capturing what you have found. You will also learn how to record or examine connections between objects as a way of either describing or theorizing about relationships in the data. Before you move on, however, do remember to save (and back up) your project!

NOTES

1 The term most commonly used in the literature is 'fracturing', but in the context of computerized coding 'slicing' seems to me to be more appropriate to describe the process of applying multiple codes to a single passage of text. Slicing suggests taking a layered view, where all are present all of the time, while fracturing suggests breaking something into separate pieces.
2 This is a very useful activity to engage in when you are struggling with conceptualization. If you aren't part of a team project, then it can be useful to link up with a group of other researchers undertaking qualitative work – the discipline area is irrelevant – and meet with them on an occasional basis. Each takes a turn to bring a short sample of data (maximum one page, de-identified) as the basis for discussion in that session.
3 In the early versions of NUD*IST, which preceded NVivo, coding was referred to as 'indexing', with nodes arranged in an Index Tree.
4 While agreeing that this may be so, Lyn also commented that "the punchline for me was the flip side, it always seemed to me that it indicated an emphasis on recurrences over unique instances and the latter may matter" (personal communication, 18th July, 2006).
5 People often ask, but it is very difficult to estimate the amount of time needed for coding. My best estimate is that, *once you have an established coding system*, you should allow at least 3 hours per hour of transcript – the actual amount will very much depend, however, on your methodological approach. This should be read in the context of understanding that experienced researchers routinely recommend allowing a working period (#days, weeks or months) for analysis of data that is two to five times as long as the period taken to gather it (e.g. Miles & Huberman, 1994).
6 Auto Coding using Paragraphs is *not* recommended for general use. It relies on each document to be coded having exactly the same number of paragraphs (including blank lines) structured in exactly the same way. Resulting nodes are identified by paragraph number only. It is designed for use only with very specific short answers which are ordered by paragraph (without headings).
7 Choosing Tools > Reports > Coding Comparison will allow you to compare the coding for two copies of the same document. This is useful (a) as a basis for discussion between team members working on the same project, or (b) where an assessment of coding reliability is required, as it provides details of the extent and location of agreement between the coders.

Chapter 5

Connecting ideas

You are ready to import more documents, and to apply the reviewing and coding processes you have learned with your first document to these new ones. Initially this may mean creating more nodes, but the number of new concepts arising in your data will drop rapidly as you continue working. As you work, you will begin to refine the ideas you are developing, and to make connections between them.

The free nodes you have been making are listed alphabetically. Some of those free nodes seem, however, to 'hang together' either because they represent similar kinds of things, or because they are related in some practical or theoretical way.

- The first way in which you will make connections between nodes is organizational rather than theoretical. It will become obvious that some of the concepts you are working with are the same 'sort of thing' – anger, frustration, satisfaction, enjoyment are all emotional responses, for example. Alternatively, some are clearly subcategories or dimensions of broader concepts – timing, cost and location are all dimensions of the issue of accessibility of services. Organizing concepts into coding hierarchies (trees) in these ways – in essence, creating a catalogue or taxonomy of your concepts – will help in clarifying what your project is about, in locating concepts for coding, and later, in identifying patterns of association between them.
- The second type of connections of which you will be aware are patterns of association in nodes. Perhaps these are nodes which you seem to use together regularly when coding, for example, involving both action and response, or attitude and experience. Perhaps they are different components of a broader concept. These patterns will contribute to your

identifying broader themes and relationships in the data. Seeing these connections is the beginning of theorizing, of moving from description to interpretation.

Until now, you have been limited to noting patterns of connection in memos as you work through your data; now you are ready to be introduced to additional tools so that, as you work though more documents, you will record and explore these connections in ways that will contribute to building and testing your emerging understanding of what is happening in your project.

In this chapter, as you introduce and work with more data:

- Organize nodes into a tree structured hierarchy to catalogue what kind of thing each is;
- Show associations in your data using sets;
- Recombine sliced or fractured data using a coding query;
- Explore the connections between concepts that seem to co-occur in your data;
- Record theoretical links using relationship nodes;
- Use see also links to show connections within and across data sources;
- Discover practical strategies for managing the coding process.

DEVELOPMENT OF THE CODING SYSTEM

Typically you will go through three major stages in developing your coding using NVivo. In the first stage, as you work through your initial sources, you create free nodes to catch ideas as they happen; this might be quick and spontaneous or slow and deliberate, depending on your approach. In the second stage, you start to sort and connect both existing and new nodes into a branching system of tree nodes that reflects the structure of the data, that is, the kinds of things that are being considered. Thirdly, you might construct meta- or more abstract codes to reflect either overarching ideas or higher order concepts, or to identify broader, more complex themes running through the data.

You will have already built up quite a number of free nodes as you have worked through the first couple of documents. As that list becomes unmanageable, or as you start to see that these nodes can be grouped, it is time to start organizing them into a hierarchical structure based on categories and subcategories.[1]

A classification system for nodes

We are all used to working with classification systems: they help us to see what we have, to understand the structure of whatever it is we're working with, and so, to identify new items and to locate existing ones. Books in libraries are classified according to their subject matter; Linnaeus developed a binomial (genus and species) classification system for plants which is still the foundation of that used today; and, possibly, your computer filing system is a tidy hierarchy of folders, subfolders and individual files organized on some principle that makes sense to you (though some I've seen defy my sense of logic!). In classification systems, the category at the top of a hierarchy describes the contents, in general terms, of the items below. Each layer in the hierarchy contains more specific types or subgroups of the item above, and may be 'parent' to more specific items below.

Already you are likely to have wanted to create a node as a subcategory of another node (compression fracture is a subcategory of orthopaedic injury), or to place it under a more general concept (anger is an emotional response). It is natural to seek to categorize and organize objects. We all create structured filing systems for our on-line documents and pictures, with folders and subfolders, just as our old fashioned filing cabinets were made up of drawers, each holding a different set of hanging files that, in turn, may have had manila folders inside, with individual documents gathered in those. In NVivo, tree nodes can be hierarchically structured with categories, subcategories and any number of sub- sub-categories, so just as folders and subfolders in your computer filing system help you to organize your files, tree nodes will allow you to organize your categories into conceptual groups and subgroups.

Think about how a department store organizes its mail-order catalogue. There will be major sections dealing with clothing, kitchen, linen, entertainment, and so on. Within the clothing section, there will be subsections for shirts, dresses, pants, underclothes, shoes. This doesn't mean that the different shirts bear any particular relation to each other, or that this dress is worn with that dress – only that those in the same section are the same kind of thing, and it is useful for clarity and comparison to have them all located in the same place. A coding system which reflects this arrangement might look as shown in Figure 5.1.
The shopper, after scanning the various sections, may then put together a coordinated

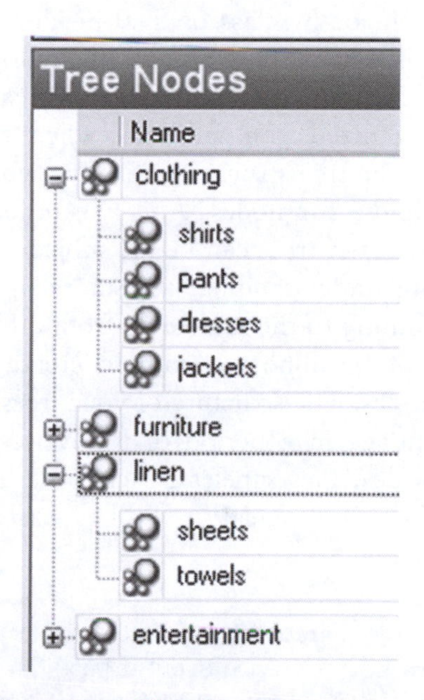

Figure 5.1 The shopping catalogue 'tree'

outfit that comprises shirt, pants, jacket and shoes. Putting together the outfit is more like making theoretical connections between the nodes: these things go together to build a larger concept, they occur together, or they impact on each other in some way. Trees are not used to show this, but rather, different strategies are used for making these kinds of connections.

Classification systems are often the subject of debate, are open to revision, and may be more or less helpful to the user, but the principles on which they are built remain. Things that are within one class are more similar to each other, with respect to critical variables, than they are to any things that are not in that class. There are at least two internationally recognized systems for classifying books (Dewey, and Library of Congress); in either, books on geography will be found in a different part of the library from books on physics, but two books on geography that are located next to each other may not be like each other in any way other than they deal with the same sub-topic within geography. In the world of Australian plants, Eucalypts will be classified separately from Banksias – not much argument there, but as botanists gained further knowledge and understanding of the development of Casuarinas, differences were perceived, and the classification of some species was changed from Casuarina to Allocasuarina. If there is a change in the critical variables around which the classification system is built, the final classification might look different, *but it will have been constructed using the same principles*. Thus the index of a recipe book might have individual recipes organized by type of ingredient or by class of final product. Which is more helpful at a particular time will depend on whether you are looking for some way of using a surplus of zucchini, or a cake for afternoon tea, but if you know the criteria by which it is organized, you will know where to look.

The tree structure in NVivo works as a filing or classification system or catalogue for nodes. And like other classification systems, the organization of nodes in trees can be the subject of debate, open to revision based on developing understanding, and more or less helpful for particular purposes. Those coming to the same data from different perspectives or with different questions will be almost certain to create differently labelled and organized trees of nodes. The structure of your trees will evolve over time, particularly when you engage in a period of review to check the consistency and salience of your nodes. Nevertheless, each node has a place with other concepts of that sort; there is a logical fit.

Elizabeth's references to her *family*, her *education* and her *work experience* were each about aspects of her *background*. Others also referred to these aspects of their lives in talking about how they came to be researching. In bringing these aspects of background together under that more general category, I am alerted to different dimensions of background experiences, possibly adding further subcategories. Again, in the same document, Elizabeth and her family were seen to view research activity in different ways: research was seen as *discovery* (by the family), and as something that *really affected people's lives* (by Elizabeth). These could be seen as two dimensions (or subcategories) of *images of research.*

NVivo uses the languages of horticulture and of the family to describe these relationships between categories. A tree has branches and, in horticulture or library index systems, the points at which it branches are termed nodes. A node and its subcategories can also be referred to as a parent with children. Thus, in the example provided above, *background* would be referred to as a *parent* node, and *education* as a *child* of that parent. *Work experience* also is a *child* of *background*, and a *sibling* of *education*. In NVivo, the way of writing the full family (hierarchical) title of education, in this case, is *background\education*. Every node is therefore identified by both its specific (hierarchical) title and its family location.

Later you will use sets, queries, relationship nodes and models to identify and record connections between concepts of different types, such as a combination of place, event and response. While these tools (described later in this chapter) are independent of trees, having your nodes effectively organized in trees greatly facilitates their use.

Why bother with trees?

If the trees don't show the relationships (theoretical connections) between nodes, then why bother with them? What is the point of just listing things in groups? Using trees to create a classification system for concepts rather than show theoretical links does bring a number of benefits, however.

- *Organization*: The trees help to create order out of randomness or chaos. The logic of the system means you can find nodes and you can see where to put new nodes. Think what kind of thing it is, and look in that tree. With rare exceptions, nodes then appear in only one place. If nodes are repeated throughout, it not only makes for many more nodes, but also for difficulty in putting together everything you know about any particular event or feeling or issue. When all you know about something is in one place, you can easily review it, and you can ask whatever question you like about it by using queries to look at it in relation to any other nodes or attributes.

- *Conceptual clarity*: Classifying codes helps to give meaning to them (Dey, 1993); sorting them into trees prompts you to clarify your ideas about what goes with what, to identify common properties, see missing categories, and sort out categories that overlap. And you will clearly see what kinds of things your project is dealing with – the structure of your data. The tree system, when established, will 'tell' your project (Richards, 2005). Indeed, when someone approaches me for assistance with a project, there are just two things I ask them to send ahead: their research question(s), and a list of nodes.
- *Prompt to code richly:* Well organized trees provide a useful tool for ensuring the thoroughness of your coding, as you progress. You stop to code a passage because an interesting issue is raised in the data. Capture that, but before you leave the passage, run an eye over your tree structure as a quick visual prompt to see if there are other nodes that are relevant. Should you also note (code) who the key players are, what the context is, how people felt or otherwise responded? Do you need reminding to always code the outcome? Making sure that the text is coded at nodes across all relevant trees will allow for much more effective and complete answers to queries.
- *Identifying patterns:* Identifying patterns of association between groups of nodes can make a significant contribution to an emergent analysis. If all events are in one tree and all responses are in another, for example, it becomes a simple matter both to see the usefulness of, and to set up a query to identify the overall pattern of which events give rise to what responses. Additionally, it will be possible to see what each of those responses might look like in relation to particular events, and so, for example, although it occurs for both, anger perhaps takes a different form depending on whether it is in response to discrimination or bullying.

Building trees

There are four basic steps to this process of creating some conceptual order in a coding system.

1. Start with a thinking–sorting process, to decide how the nodes might be arranged. When you ask yourself why you are interested in a particular category or concept (abstracting), ask yourself also what sort of thing it is (classification). Work through each of your free nodes in this way, making a list of the sorts of things that you have there (if you're struggling to do this, you might find the section on *Kinds of trees*, below, will give you some ideas).
2. Create top level nodes as needed. You may already have some free nodes which are broad categories (e.g., culture, attitudes, environmental issues), suitable as top-level nodes – these can be moved directly into that

position. Others will be specially created for the purpose of providing a structure, rather like coathangers for others to hang from.

3. Move free nodes into trees, so their order reflects that arrangement. Not all nodes will immediately find a place in the system – that's OK. Some free nodes may need to be copied into two trees because they were actually about two things, in which case, they will also need renaming once they are in their new locations.
4. When you're done with shifting nodes and you have a structure, check that it serves the main ideas you set out to work with and the research questions you want to answer. If there's a tree that doesn't fit, ask whether you need to modify it or your original ideas. And when you go to create a new node now, you should be able to see where it will go.

Regard your tree structure as a work in progress. Some nodes may take longer to place (and may never be), and others will move between trees until you crystallize just what it is they are 'a kind of'. Some start out as a fairly random collection under a broadly-termed tree that you return to later to sort. I often have an *issues* tree (for things about which there might be debate for this topic) which later turns into issues of various types as I gradually realize that I am dealing, say, with structural issues, political issues, interpersonal issues, and so on; these are then set up as branches under the top level node of issues, with the specific issues under the relevant branches.

The secret to success with arranging and using nodes in trees can be summed up in the following 'rules of thumb' which have been crystallized over many years of researchers' experience in working with hierarchical coding systems (see also Richards, 2005, Chapter 6):

- Organize trees based on conceptual relationships (the same 'sorts of things'), not observed or theoretical associations;
- Use a separate node for each element (who, what, how, when, etc.) of what the text is about;
- Each node should encompass one concept only;
- Each concept appears in only one tree in the whole system;
- A particular passage of text will be coded at multiple nodes; and
- Keep the system 'light'; be flexible.

When I initially started working with some data about what motivated researchers (in an early attempt to use NUDIST – version 2!) I created a node called motivation and then created a series of nodes under that of things that had the effect of motivating academics to do research. That was fine until I began wondering what to do with these things if they were talked about in a different way – perhaps as part of the researcher's developmental experience, or simply as an outcome of their work, or even as a limitation. And so they reappeared in trees for development, and/or experience, and so on – and unfortunately not in neat repetitions. After some period of struggling with this cumbersome system, application of a flash of insight saw what had become a four page list reduced to one: have a tree for impact (with subcategories of stimulate, develop, maintain, reinforce, limit), with separate trees for each of the kinds of things that had an impact, such as people, events, activities, and other trees for the contexts that moderated those things, and for the way people felt or otherwise responded. For any interesting segment of text, code at as many of these as were relevant, and always think about applying a code for impact. This would allow me to rapidly determine the impact of any person, event or activity, and to see when it is that something stimulates research, and when it becomes a limitation, or to find all those things that might act as stimulants to research motivation. The difficulty I had then was that re-organization of the existing nodes was complicated by their varied organization and usage. Coupled with a sense that a lot of what was in the data had been missed because of the way I had been coding it, this led to my scrapping my first coding (and its attempted reorganization) and starting again. Early intervention is a much better option!

Kinds of trees

It is perhaps surprising to new researchers that it is possible, ahead of time, to predict what will be some of the kinds of trees you will use in your study. For research projects which deal with the lives and interactions of people, this is broadly possible because the kinds of labels that most appropriately go at the tops of trees are typically quite general terms. This is not to say that you can't have more specific types of trees, nor to suggest that these are mandatory trees. It might be helpful to those struggling with the idea of trees, however, to indicate what kinds of things might be separated out into distinct trees, and to provide some sample coding systems from real projects.

Typically I would have just one, and sometimes two 'generations' of nodes below a top-level tree node. Miles and Huberman (1994: 61) also suggest that "many researchers use a simple two-level scheme: a more general 'etic' level … and a more specific 'emic' level, close to participant's categories but nested in the etic codes." The kinds of trees which I find turning up again and again in projects involving interviews with people include (but are not limited to) selections from the following list. The list is far from exhaustive – it is provided simply as a stimulus for thinking. Your list will be influenced by both methodological approach and discipline, as well as by the substantive topic you are investigating.

- *People/Actors/Players* – people, groups or organizations to whom reference is made. These are rarely, if ever, coded without coding also (on the same text) the reason for the reference to that person, organization or group. Depending on the project, these nodes might be very specific (Dr Smith), or simply role based (doctor, nurse, manager, partner, friend).
- *Events* – things that happen at a point in time.

- *Actions* – things that are done at a point in time. These would usually be coded also by an Actor node (unless it is an action of the speaker).
- *Activities* – ongoing actions.
- *Context* – the settings in which actions etc. occur. This may include branches to identify phases, stages, timing, location. (Note that if contextual factors apply to whole cases, attributes should be used rather than nodes.)
- *Strategies* – purposeful activity to achieve a goal or deal with an issue. In some projects this might be more specific, as coping strategies.
- *Issues* – matters raised about which there may be some debate. Typically both 'sides' of the debate are included under a single node. If the number of these proliferates they might later be grouped into types of issues, e.g. political, structural, interpersonal, personal, environmental, with each of these being represented by branches, or perhaps as separate top level trees.
- *Attitudes* – listing the type of attitude, rather than the focus of the attitude (that gets coded separately).
- *Beliefs/Ideological position/Frameworks* – intellectual positions (or discourses) which are evident in thinking and action
- *Culture* – likely to have a number of branches, depending on the type of culture (organizational, societal, etc.) being considered.
- *Emotional responses* (feelings).
- *Personal characteristics* – descriptors of the person.
- *Impact/Outcomes* – e.g. facilitator/barrier, or help/hinder, etc. In general there should not be further nodes under these specifying particular facilitators or barriers, rather they should be double-coded with whatever is acting as the facilitator or barrier.

As well as the types of nodes suggested above, which are all directed toward capturing the content of what people are talking about, there is at least one other area worth considering:

- *Narrative* – to pick up on the narrative features discussed in Chapter 4, such as turning points, omissions, contradictions, high (or low) emotion, climax, objectification, subjectivity, and so on. This could be used also as a place to locate *suggestions*, or *good quotes*: these would not have additional nodes under them, but the text coded there would always be double-coded with what the suggestion or the good quote was about.

Lyn Richards also suggests having an area for 'retired' nodes, in case you're too scared to throw them out. What all this points to is that you should bend and use the coding system to suit your own purposes.

You will find some sample coding systems (from real projects) on my website (www.researchsupport.com.au), on the resources page for this book – these might provide you with further clarification or ideas. Additionally, if you already have a coding system which is in a tangle, with a multiplicative (repeating) node system, you will find suggestions there for ways to sort out the mess. NVivo **Help** offers guidance under **Making Nodes > Building Efficient Node Hierarchies.**

REARRANGING NODES

Before making drastic changes to your project, be sure to save it. Also, if you haven't done so recently, it would be a good idea to close it, and make a backup copy (in Windows – making sure the project is closed when you do), then reopen it to start working on your new arrangement.

Creating a list of nodes

This is for those who need to work on paper first!

- Create a **Node Summary Report** for your free nodes, including information about how much each has been used, from the **Tools > Reports** menu; OR
- Print a simple list of free nodes: From the List View of your free nodes select **RMB > Print > Print List**; OR
- Export a list of free nodes in table format: From the List View of your free nodes select **RMB > Export > Export List**; choose whether you want to save the list in Excel or Word format, and where you want to save it; then open (or print) from that location.

Use the modeler to help think about which nodes are the same sort of thing and therefore belong in the same tree: In a new model, **RMB > Add Project Items > Free Nodes.** Do not add any associated items (so just click OK). As you are thinking about what goes with what, if you need to see the text to remind you of what a node is about, select it and **RMB > Open Project Item.**

Once you have worked out which nodes you want in which trees, it is time to set up the structure, move nodes into it, and then make use of it.

Create a parent tree node

- Go to the **New** button on the Main toolbar (top left) and choose to make a new **Tree Node** (from any view); OR: Select **Tree Nodes** in Navigation View. In the List View area, *below any existing nodes*, **RMB > New Tree Node (Ctrl+Shift+A).**

- Provide a name for the new node (and a description if you want).

OR

- Drag selected free node/s across to **Tree Nodes** in Navigation View. This will place all selected nodes at the top level; they can then become parent nodes, or be dragged from there into particular trees.

Moving from Free to Tree

- In List View select one or more free nodes. **RMB > Cut** the selected nodes; click on **Tree Nodes** in Navigation View to change the List View display; select the appropriate parent node for the nodes you have selected; and **RMB > Paste.**

✓ If a node needs to go into more than one tree, use **Copy** then **Paste** into the first tree, then return and **Cut** to paste into the second tree. Rename the node appropriately in each new location. The text at that node will now be coded in two places.

Moving nodes in trees

- Nodes can be dragged from one tree to another. Alternatively, **Cut** *or* **Copy** and **Paste.** Dragging or paste will place the node under the node you drag to or paste at (so you are moving it to a new parent). To place a node at top level, drag to or paste at **Tree Nodes.**

✓ If you are having trouble dragging a node from one tree to another (dragging just seems to select everything in between), make sure you first select the node, *then* click on it to drag it.

Merging nodes

- If you have two nodes which are about the same thing, then **Copy** the first node (or **Cut** if you are confident), select the second, and **Merge Into Selected Node.** This will place all text references from the first (source) node into the second (target) node.

✓ I usually amend the description of the target node to indicate that I have merged another node with it, by adding "includes [source node]".

✓ You can select more than one source node at one time, to merge with a target node. Check the RMB options for alternatives if the node has children.

CODING WITH TREE NODES

- Code with tree nodes as you would with free nodes. If you are using drag-and-drop coding with the tree nodes displayed, remember to change the display to Detail View Right, and to turn on the coding density bar to prevent the text being accidentally moved.
- To create a new tree node as you are coding: right-click on the parent node, then select **New Tree Node**.
- To expand a tree, click the + next to the top level node (double clicking will open the text for that tree node).

✓ If you 'lose' a node because you can't remember which tree it is in, you can locate it using **Find** in the bar at the top of the List View.

Keeping an audit trail of changes

It is times like this, when you are making significant changes to your coding system, where keeping a record of the changes you are making and why you are making them is particularly important. In the process of making decisions about your tree structure, you will be thinking seriously about the goals of your project, and the concerns it embraces. Keep a record of these thoughts in your project journal (it can be found in the Memos folder, in Sources) as they will assist when you come to writing up your project.

Additionally, you will find it valuable to store (dated) lists of nodes at various points during your project. These provide a historical archive which helps to chart the shifts and progress of your thinking.

MAKING CONNECTIONS ACROSS TREES (AND OTHER ITEMS)

Having sorted (classified) nodes into trees, we will now turn our attention to the more theoretical kinds of connections you might see and want to make between nodes (and other project items). "Connecting concepts is the analytic equivalent of putting mortar between the building blocks" (Dey, 1993: 47).

Miles and Huberman (1994: 57–8, 69–72) use the term pattern coding to describe a second level of coding which is more inferential and explanatory, and which is likely to be applied later in the analysis process when the significance of particular comments or observational notes is more evident to the researcher. What they are describing as pattern codes would typically encompass or connect concepts that come from any two or more of your trees. They propose that pattern codes are generally of four types: themes, causes or explanations, relationships, and emerging constructs; and that they serve to refine categories, prompt early analysis, focus data collection, build a conceptual framework in which to place incidents and interactions, and to lay the groundwork for cross-case analysis. They note that the "most promising" pattern codes should have memos attached to them, and finally, they suggest trying to map (model) these codes to help see their interconnections and so develop the conceptual framework for the study. (In general, I will use the term *metacode* rather than pattern code to avoid possible confusion, at a later stage, with pattern analysis.)

At the risk of pushing the shopping catalogue analogy too far: think about the ways in which various separate items (a shirt, a pair of pants, jacket, shoes, etc.) might be put together to make an outfit. These might be brought together as a coordinated outfit (together they make for a larger concept); you might choose a combination because it's something you've seen together often (they regularly co-occur); you could choose a particular scarf to go with this blouse because the colouring of the scarf will bring out (i.e., impact on) the highlights in the pattern on the blouse; or, having bought a particular dress, you go looking for a matching pair of shoes. In NVivo, you can generate and/or record these types of connections through use of sets, coding queries, relationship nodes, see also links and models.

Clustering items in sets

The term axial coding, used in grounded theory, is designed to express the idea of a cluster of concepts (the spokes) which are focused around a central idea (the hub, or axis) (Strauss, 1987). Axial coding is seen as being an intermediate level of working with concepts, somewhere between open coding (a kind of free-for-all flurry of ideas prompted by the data, typically recorded in free nodes) and the development of a theoretically rich core category which identifies a central process in what is being observed. Strauss and Corbin (1998) talk about axial coding as reassembling fractured data. Often axial coding will bring together different elements from the coding paradigm – a particular set of conditions, strategies and consequences. Being able to cluster nodes in a set, which is itself named with a broader concept, meets this need of being able to identify nodes which have some meaningful relationship together. And it does this in a way that takes you to a higher level of abstraction, again helping you to move beyond description into theorizing.

Other methodological approaches often employ a similar strategy, although they may use different terms. Phenomenologists write about creating potentially overlapping clusters of units of meaning to form themes, or units of significance (e.g., Hycner, 1999; Moustakas, 1994), and may engage in doing so on a case by case basis as they progress through the study (Smith & Osborn, 2003; *cf.* Chapter 8). In this process redundancies are eliminated and the meanings of the various clusters are interrogated in order to identify 'essential' themes. Pattern codes or metacodes are terms used by others to refer to material pulled together into more meaningful and parsimonious units of analysis, somewhat analogous to cluster or factor analysis (Ryan & Bernard, 2003). In theory-emergent methodologies, the number of codes-in-use may decline as thinking about the issues in the data shifts from noting the descriptive or particular to a more comprehensive, generalized or abstract process.

In NVivo, you can use a ***set*** to flexibly group items (sources or nodes) in the project. This is the simplest way of showing that items are connected, and giving that more abstract or global concept or theme a name. Sets are therefore useful for grouping together nodes that focus around a common, broader concept, or are connected in a broader theme or theoretical relationship. These items will often consist of different 'kinds of things' that would not be in the same coding tree. While sets can also be used purely as an organizational tool (e.g., to gather all the first wave of interviews), in the current context of making connections they are doing more than that, and in naming a set of this kind you are effectively creating a pattern or metacode, or identifying a theme.

> When I began to see *obsession* as a critical concept in describing researchers, and to understand obsession as a kind of driven passion, I looked for nodes that might be capturing text related to that broader concept. These included some that had been listed as personality characteristics (e.g., commitment) and some from the experience tree (e.g., addiction, excitement, enjoyment). Putting those into a set served as an indicator of the broader concept, and provided the means to explore and refine the concept further as work developed. Did I need to now distinguish between general enjoyment and responses which might better reflect the higher level of passion, for example?

To create a set of items in NVivo, you provide a name for the set, and then identify those items that belong in it. In so doing, you are not changing the items, nor are you moving them or duplicating them. You are simply saying that these items belong together in some way (as expressed in the name for the set). What happens in the software is that an alias for each of the items is placed in the set; these are seen on the screen as 'shortcuts' to the items. This means that items can belong to more than one set. Placing items in a set has at least five benefits in your project:

- The visual evidence of seeing items together helps you to see their connectedness (you might record a comment in your journal about this).

- A set can be placed in a model with an option to immediately show its members (as associated items).
- A set can be used as the basis for selecting coding stripes to be shown alongside the detail view of a document or node.
- A set can be incorporated in a query as a single unit of data, thus allowing the broader concept to be used in asking questions of the data.
- A set can be saved as a node, allowing all data relevant to a broader concept to be brought together, refined, and used in further analyses.

CREATING AND POPULATING SETS

Any kind of documents or nodes can be added to a set as members.

- ▶ In any Node or Document List View, select one or more items and **RMB > Create as Set**. Name the new set. Items can be members in more than one set.
- ▶ If the set already exists and you wish to add to it, select one or more items from any List View of nodes or documents, then **RMB > Add to Set**.
- ▶ To view the set, choose to show and expand Sets in the Navigation View. Members of a selected set will show in List View as aliases.
- ▶ Modifying a set: deleting an item from a set will not delete the item, just the shortcut to it. If you open an item from a set and modify it, however, you are modifying the actual item.

'Trust me!' Seeing connections in sliced or fractured data

Separating out the component categories or concepts conveyed by a passage of text, as recommended in Chapter 4 and as reflected in the structure of node trees, presumes a capacity to put them back together again. This is the task of the coding query, where it is possible to ask for text which is coded with, say, *encouragement* AND *supervisor* to find just that text which is about how supervisors provided encouragement to engage in research. The coding query is the antidote for sliced or fractured data! The query process, for this kind of question, draws on your having coded the same text at more than one node; it finds all text satisfying the condition that it is referenced by (coded at) both nodes – often referred to as an *intersection* of coding (Figure 5.2). In knowing the software can find these connections, you can be more confident about coding at separate nodes each component of what is being discussed or told in your data. As you move into using trees to organize your nodes, such confidence becomes even more vital for an efficient coding strategy.

Finding intersecting coding is very easy for a computer to do; it would be much more difficult for you to do this accurately without a computer. The rigorous sorting and matching capacity of the computer for tasks of this kind is one of the reasons why using a computer for analysis can involve different coding strategies than coding for manual analysis.

When you run a query, the text that satisfies the criteria you set in the query is found. It is then *your* task to read and interpret this text – there is no equivalent of a statistic with a *p*-value in qualitative analysis! The text can be previewed, in which case it isn't saved after you close it, or it can be saved, in which case a node holding the results will be created in the Results area of the Queries screen, for immediate or later viewing. Results nodes do not automatically update when you add and code more data in your project, but if the query has been saved, you can re-run it to obtain updated results.

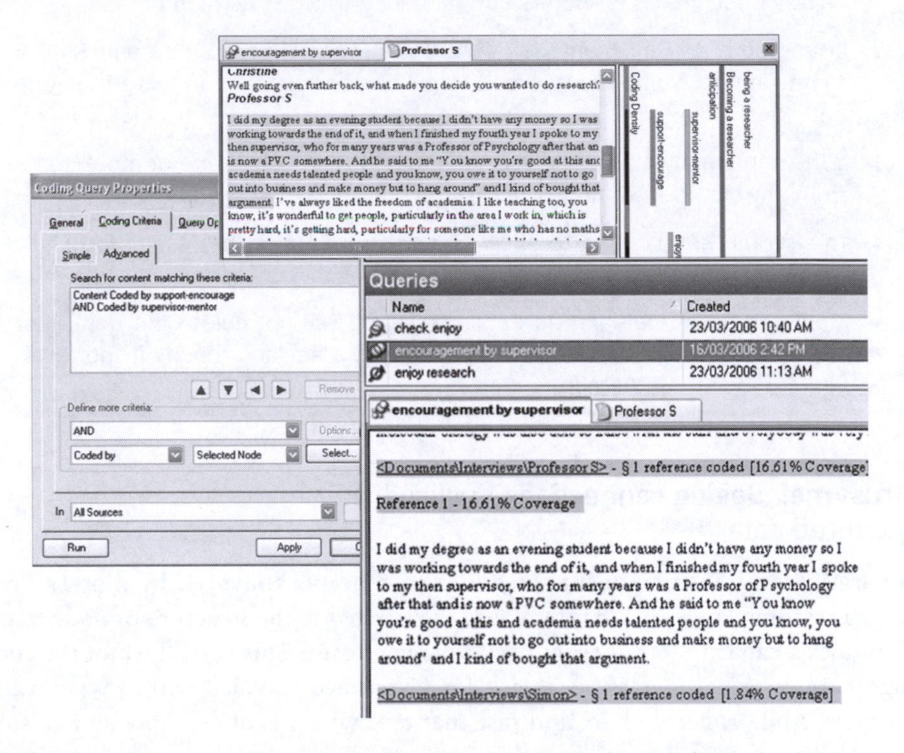

Figure 5.2 Using a coding query to reconnect (intersect) coding

A FIRST CODING QUERY

- Select Queries in the Navigation View, then right-click in the List View to create a **New Query**. You want to create a **Coding** query.

- To save your query so you can run it again later, check next to **Add to Project**, and provide a name for the query.
- Return to **Coding Criteria**, and select the **Advanced** tab. Don't be put off by the sound of 'advanced' – it just means you are going to be using more than one node!
- Select the nodes to add to the query by clicking on the **Select** button. You will be shown a list of your nodes. Once you have found and checked the nodes you want to use, click **OK** (or press Enter on your keyboard). To complete the selection, you now need to click on **Add to List**. The nodes will be entered into the query dialogue.
- Now check the **Query Options** tab. You will find the query is set to show a **Preview Only**, which is all you need for the present.
- Click on **Run** at the base of the dialogue. This will both save the query (if you elected to do that) and run it. (Clicking OK will simply save it.) The results of your query will open in Detail View. If you decide, on viewing the results, that you want to keep them after all, you can right-click and choose to **Store Query Results**.

For further information, see **About Queries**, and **Advanced Coding Queries** in **Help**.

Queries in NVivo can be quite straightforward, such as the present task of asking for all instances of where text is coded by this node AND that node, or they can be made quite complex, with multiple restrictions on what is found, how the items might be associated, and where the search is conducted.

Now that you have used a query for checking connections where you had purposely used two nodes rather than one to code a passage, and you have seen what it can do, try using the same type of query to explore other associations about which you might be developing some hunches because the codes seem to co-occur quite often, or the concepts seem to be associated. From quite early in the Researchers project, for example, it seemed that enjoyment of research was often associated with the stimulation that research provided. This could be checked out with an advanced coding query – first of all using AND, but then also by trying it with NEAR (within the same scope item, with the data scoped to interviews), because both elements were not always spoken about at exactly the same time.

As you gain more experience with NVivo, you will find yourself using queries throughout your project: they are *not* something to be left until the last stage of analysis.

Identifying and coding relationships between items

It is important in understanding patterns to identify not only the concepts, but the *linkages* (or relationships) between them. The hierarchical ordering of concepts in a tree-structured coding system – as categories with sub-categories, or concepts with their unique dimensions – does not directly show the associations you find between different kinds of things. Nor is it easy, in a hierarchical classification system, to capture information like who communicates with whom about what. These kinds of connections – the kinds which build theories rather than concepts – need a different solution, and one that is offered here is to capture them in relationship nodes.

Relationship nodes record a connection of a particular kind between two project items. The project items might include cases, categories or concepts (in nodes); documents, memos or externals; or sets of objects (treated as a single unit). The relationship can be a simple association, or it can have one-way or two-way (symmetrical) directionality (represented by a line, a single-headed arrow and a bi-directional arrow respectively). The particular type of relationship is recorded also, for example, *encourages, impacts on, communicates with, loves*. Thus, using a relationship node, you can record *friend encourages Elizabeth* as a directional relationship, or *Ange works with Paul* as an associative relationship, or *frustration leads to violence*, also as a directional relationship.

Because relationships are held as nodes in the NVivo system, you might also choose to code supporting evidence for the relationship, either as you are working through your data, or from the results of a query. If you are recording something like who communicates with whom, and coding instances where this occurs, then as for any other node, you might also code the same text with the subject of the communication, or the means of communication (using tree nodes). You will then be able to examine the pattern of who communicates with whom about what, and in what way. (You might also create a relationship node for *does not communicate with*!) Additionally, relationships can be viewed in a model, where related items can be brought in as associated items, for example, to show all the sources of support and encouragement for Elizabeth, or all the people with whom the CEO communicates, or all the events which might trigger a personal breakdown.

There are two stages to setting up a relationship node. In the first, you will classify what kind of relationship you are dealing with. This includes both its directionality, and its specific type. Then, in the second, you will apply that classification to a particular situation by specifying who or what project items are related. Once created, it will be listed under Nodes, in the Relationships area, and will be available for coding or including in queries or models, like any other node (Figure 5.3).

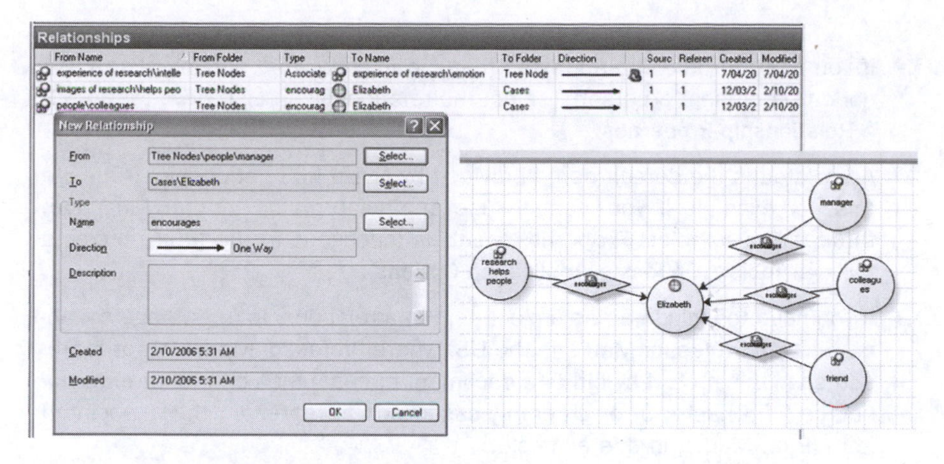

Figure 5.3 Recording and modeling relationships

RECORDING RELATIONSHIPS

Setting up the relationship type

- From within the **Classifications** area in the Navigation View, select **Relationship Types.** In the List View, right-click to create a **New Relationship Type.**
- Provide a **Name** for the relationship type you are creating (e.g., encourages, works with, talks to, impacts on). Enter a **Description** if needed.
- Select the **Direction** for this type of relationship, using the drop-down options. Click OK.

Recording a relationship

- Move to the **Nodes** area, and select **Relationships** in the Navigation View.
- Right-click in the List Area to create a **New Relationship**. Make the necessary selections for the source (From) and target (To) of the relationship, and the type (Name).
- The new relationship node will appear in List View. It can be coded and viewed like any other node.

For reasons which will become apparent later (*cf.* Chapter 6, dealing with groups based on attributes in models) it is advisable to use case nodes rather than documents (where appropriate) to represent research participants as part of a relationship.

- If you need to edit a relationship (e.g., to change from document to case node to represent a person), select the relationship in List View, then **RMB > Relationship Properties.**
- **Associated** is the default name provided by NVivo for relationships. Change this, for example if you want to use another term or another language, through **File > Project Properties > Labels** (change it for all *future* projects through **Tools > Options > Application Options**).
- If you need to include a non-project item in a relationship (e.g., Ange collaborates with Professor Marks in the US, where Professor Marks is not represented in this project by either a document or case) then create an ***external*** source (*cf.* Chapter 3) or an empty case node to represent the associated item (Prof. Marks, in this example).

See also links and hyperlinks – connecting a web of data

Working with data is not just about coding. Immersion in the data sparks recollections of things read elsewhere. ***See also links*** make it possible to link from a point in the text to another document, or to a selected passage in another document (*cf.* Table 4.1). In so doing, they can meet a variety of needs:

- Link from a passage to notes from reference material dealing with the issue raised in the passage (Figure 5.4).

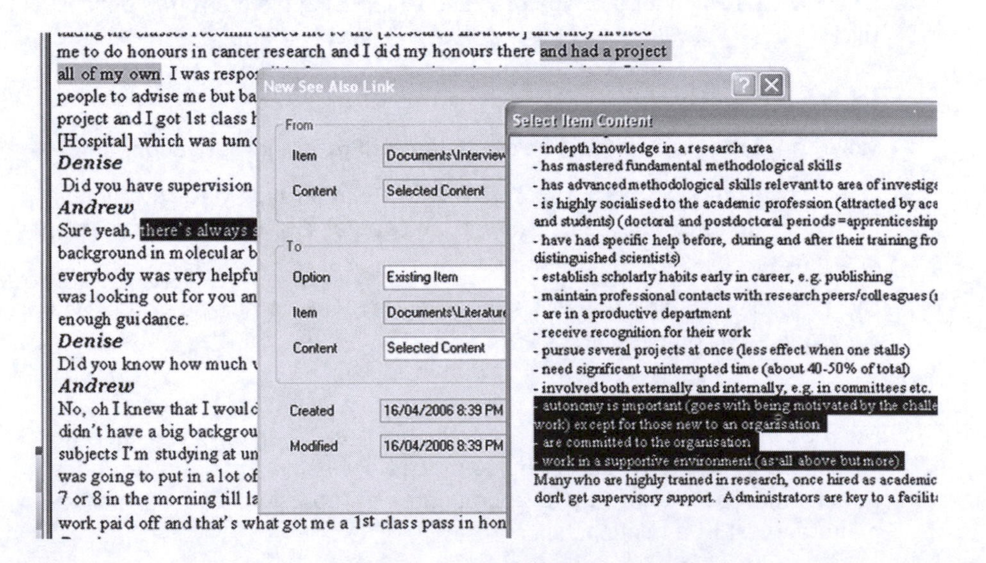

Figure 5.4 Creating a see also link to selected content

- Link from one passage in a document or case to another in the same document or case where the one provides an explanation for or expansion of the other, or perhaps a contradiction of the other, for example, to link witness accounts in a legal case.
- Link from a passage to a note in a memo prompted by that specific passage. You've thought long and hard about the issue this person has raised, and it's too much for an annotation – record it in a document or node memo, but then place a see also link to that particular reflective comment within the memo. Its roots in that passage will be evident and you will be pointed directly to the linked thoughts (rather than to the memo in general) whenever you see that passage.
- Link across documents to build a sequentially ordered picture of an event through the eyes of different tellers, or to trace an evolving idea or saga (see the story of *The outback governess*). When a see also link is accessed (**RMB > Links > Open to Item**) the linked item is opened with the selected passage highlighted, thus allowing a further link in the web to be created.

One important advantage of the option to link to selected text is that the text of the linked passage can be printed as an endnote on any document or node report which includes its anchor.

The 'outback' governess

In her Bush Schooling tutorial for NVivo, Leonie Daws used links to trace, through a series of 7 emails to an on-line forum, the story of a governess' coming to, and rapid departure from, a property in outback Australia. See also links allow the story to be told in a sequence that is lost when data are simply gathered at a node.

In her first message (1), Evelyn writes to her on-line network expressing concern about the prospect of appointing a governess to help with 'School of the Air' for three young children on the property, and seeking the wisdom of the other women. She writes next (2) of becoming even more concerned about how the governess would cope, after meeting her for the first time. Her following communication to the forum (3) told of what happened when the governess finally arrived: "We have just had our first inaugural experience with a governess. She came … and she went! Didn't even start school!" Evelyn wrote later (4) that "one of the reasons the governess didn't work out (besides the fact she was having a nervous breakdown) was that she became rather fond of my 'attractive' 'caring' husband (her words … he is all these things BUT he is mine!!!!) in the 4 days she was here." Other mothers rallied to her cry of desperation (5) with wry comment ("governesses [are] either chasing a hectic social life or your husband") and moral support (6), telling stories of both disasters and wonderful friendships made: "Just don't mention the year when six passed through the house in one year. I asked myself all the same questions! Thank goodness after that we had a wonderful young woman stay for 18 months, and the next one stayed for 4 years and both have become great friends." Evelyn concludes the story, shortly after, with some good news (7): "Just thought I would let all the wonderful people know who gave me recent support in my governess saga, that YES someone IS looking after me!!!!! Last night an experienced teacher moved in. She is, as I write, buried in the schoolroom, keen as mustard..." and "I am so lucky to be able to access such 'been there done that wisdom'!!!!!"

Perhaps your reference material exists only as journal downloads in .pdf format; the group discussion links to an active web site which is constantly updating; or the emotion in the distraught mother's voice is best understood through a 'sound byte' from the original tape. For such situations, ***hyperlinks*** similarly allow you to make direct connections from a project item to other data, except that these link to on-line data or illustrative material which is not directly part of the project, such as journal articles, web pages, video or audio files. Use hyperlinks to link from your journal to records of meeting with supervisors, emails from colleagues, and other sources of influence on your developing thinking as well, as part of an effective audit trail for the project (Bringer *et al.*, 2004).

See also links can be created from within either a document or a node, but hyperlinks can be created only from text in an editable document. Both, like annotations, travel with their specially marked text anchor into nodes and results of queries, and can be accessed equally well from a document or node.

Using the modeler to map connections

Previously you have used the modeler to map the conceptual framework for your planned project (Chapter 2), to visualize a particular case (Chapter 4), and perhaps, to map relationships. Now, as well as continuing to review each case using a model, you can use the modeler as a tool to visualize conceptual and theoretical connections in your data. Create a new model and add one or more project items, with or without associated items, then adjust the display to tell the story of your research.

- Organize categories around a central explanatory concept.
- Map specific connections between nodes, to reflect processes or associations you have noticed in your data.
- Map your more inclusive codes (axial-, pattern-, metacodes – possibly using Set icons at this stage) to develop a revised conceptual framework.

Use model ***styles*** (*cf.* Chapter 4) to help communicate what you are beginning to see in the data. Then write about what you have found in your project journal.

CODING IN PRACTICE

Knowing about how to code in theory, and making it work effectively and efficiently for your project can be two quite different things! What follows is a series of pointers to things you might need to consider, practical issues to be aware of, and additional hints to help you along the way. Many of these issues have come up regularly as questions or concerns of participants in workshops, or they are lessons from my experience in dealing with client projects.

Do I code the question?

When a conversation is being held about a particular topic, there is often an exchange between the participant and the interviewer while on the same topic area, such that you will want to code more than one paragraph or speaking turn. When you do so, include the interviewer's intervening text as well for methodological and practical reasons:

- it is helpful to know how the interviewer was prompting or responding to what the participant was saying, and
- every time the coded passage is broken by a non-coded interviewer's turn, then the parts will be displayed as separate retrievals and the text will consequently be broken up when the node is being reviewed.

Where the passage you are coding is not interrupted by the interviewer, you may have no need to include the interviewer's preceding question or comment, but if the question is needed to understand the participant's response, then it most certainly should be included. Otherwise, any time you are reviewing text for that node, you would have to return to the original document in order to see what the coded text meant.

Code at the parent too?

When you code at a child tree node (subcategory), the coding references are stored at that node only; the text is *not* automatically coded at the parent tree node (larger category) also – and nor, in general, would you want it to be. The parent node might be used to store general references to that concept, or as a temporary dumping spot for text that is still to be coded to more detailed child nodes. It may also serve as the location for a linked memo dealing with the general category of things that tree (or branch) is about.

Occasionally there are situations, however, where you *do* want text to be coded at both broad and specific levels (parent and child) for the same concept. For example, in a project analyzing injury types related to compensation for motor vehicle accidents, the team needed to consider injuries at both general and specific levels; thus orthopaedic was parent to various types of fracture and dislocation; psychological to anxiety and depression, and so on. Most analyses in the injury study were conducted at the broader injury level, but the detail remained available should it have been needed.

It is quite a simple matter to gather text coded at a set of child nodes, and *copy* all of it into the parent node. In List View for Nodes you can select multiple nodes (at the same level) at the one time, so a set of children can be selected together (use click-shift-click from first to last to get all at once), copied as a group (**RMB > Copy**), and then merged with the parent node (**RMB > Merge Into Selected Node**). Note that this is probably one occasion where you do *not* want to copy all the associated items (memos, see also links, relationships) when merging nodes. If you repeat the merging process later in the project, because you have done further coding at the child nodes, that will not be a problem – text will continue to appear once only at the parent node, and meanwhile it will remain also at the child node.

How many codes in a project

New nodes proliferate early in the project, but as concepts and themes recur in further documents, you are more likely to add to existing nodes. If you haven't added further data to a node by the time you have coded a few documents, then it will be time to rethink. The categories may be too specific to apply to more than one document, and if that is the case, they tend to be of little value for analysis. You might run a check after a few documents to identify those nodes that code only one passage (number of references), then challenge each of these with Lyn Richards' question: "Why am I interested in that?" (*cf.* Chapter 4, also *reviewing nodes* in Chapter 7) to identify a broader concept or purpose for this node. The discipline of writing a description for each node (in the Node Properties dialogue) can also help sort out what they are about, and which you need and which you don't.

Experience teaches that projects typically don't have more than about ten trees, and that trees usually don't go more than two or three layers deep; it just isn't possible to subcategorize much more than that without starting to confuse what class of thing you are dealing with. There may be a large variation in the number of nodes within a tree, although if you get too many (say, 30+) at a single level in a tree, it is probably time to group them into further branches.

Coding with analysis in view

Coding can be of a word, phrase, sentence, paragraph, long passage, or a whole document; the length of the passage to be coded is totally dependent on context and analytic purpose. Knowing how queries have been programmed to work in the software, however, means that there are some considerations that you are best aware of from this early stage. Generally speaking, you need to include a sufficient length of passage for the coded segment to 'make sense' when you retrieve it, but at the same time, avoid coding long, drawn-out, vaguely relevant passages that are going to 'clutter' your reading when you review the text at the node. This is about balancing specificity and context.

Coding very small segments (say, a word or a phrase) can create retrieval problems, and difficulties in using those nodes in queries:

- Very precise coding of multiple small segments within the same broad passage at the *same* node will result in multiple retrievals rather than the single one that would result from coding the whole passage at the node. Not only is this hard to read, it stretches the memory of the computer because retrieving 200 passages is much more demanding than retrieving 50, even if there is less actual text retrieved.
- Sometimes, rather than coding a whole passage for all nodes, researchers code each tiny component of the text at separate nodes, and so connections between nodes become less clearly defined. Rather than being able to use AND queries to find associations, the researcher has to rely on NEAR queries, where it is much more difficult to specify what text should be included. The results from these require much more careful evaluation, as there is less guarantee that the association is meaningful.

Design your approach to applying codes (as for everything else) with analysis in mind. It helps, here, to have a basic understanding of how the reporting and query functions of the software work. This is rather a 'big ask' at this stage, so here are guidelines to help with a couple of areas most likely to create problems:

- If you find you want to add further coding immediately following the passage you have just coded, select the text to ensure the passages overlap so that the two are counted together as one when NVivo gives counts of coded passages for that node (unless, of course, you really do want it to be counted as two). Also, it will show in the detail view for the node as a single retrieval.
- If you want to find whether things that are done or happen are impacted by the context in which they happen, then be sure to double-code any text being coded for actions, strategies, events or whatever also with a node for context, even if the context is not specifically mentioned in that part of the passage (as long as it is clearly relevant). Coding context as well as strategy on the same passage facilitates discovery of patterns of association between strategies and contexts using a matrix coding query. Similarly, if you want to examine responses (emotional, political, physical, etc.) to issues, events, decisions, behaviours or other stimuli, ensure

passages are coded for both elements. For example, if you are asking about the strategies that counsellors adopt to deal with adolescent behavioural problems, then the description of a particular strategy, if used for a particular problem, is best coded for both problem and strategy, even if the text referring to the problem is in an earlier part of the passage.

Flexibility and stability

> ... codes are organizing principles that are not set in stone. They are our own creations, in that we identify and select them ourselves. They are tools to think with. They can be expanded, changed, or scrapped altogether as our ideas develop through repeated interactions with the data. (Coffey & Atkinson, 1996: 32)

Particularly, but not only, where you are creating categories out of the data, expect that your ideas about what is relevant or important to code will develop and possibly change. It is very easy to become beguiled by a thread of thinking that you later realize is not important or relevant to the current project, or to have codes that you set up early, but then find are a struggle to use, and which finally don't advance your understanding of your topic. At the same time, setting the parameters of what might be relevant early in the project and limiting change thereafter is deadening. Beware 'shoving' things into a node – think about whether they actually fit. Check what's already coded there, and if you're not sure, create a new node; it can always be dropped or merged later if further data for it isn't forthcoming.

One of the most telling experiences I have had, relating to the deadening impact of 'freezing' nodes, came from a government employee who had laminated her node list (a printed list was especially useful in the days of NUD*IST, as a basis for coding). She was determined, after having made some initial adjustments, to make no further changes to her list and therefore to what she was looking for in her data – and she was 'bored witless' with the process of coding (and the project, which was never completed)! Your coding system will stabilize, but it should nevertheless remain open and flexible, and coding should be interspersed with other activities (memo-writing, querying, modeling) in order to keep your thinking focused and fresh.

Perhaps you fear losing information, or having unreliable nodes, if you change your coding system as you work? Flexibility about ideas of what is important to the project and about how to arrange those ideas is needed, but so is some stability of purpose or sense of direction to avoid being tossed about by every new idea arising from the data. Consistency in what you are calling things is also necessary, to be able to make use of the final set of nodes. The issue, then, is to balance flexibility in coding with purposefulness and consistency.

Use the Description field in the Node Properties dialogue to record what a node is about, including perhaps a record of how the concept was developed. Alternatively, if the history is significant, record its development in a linked memo.

In the Researchers project, just a few documents in, I started to develop a series of codes dealing with career development, and went back over earlier documents to find text I'd missed on this topic before realizing that that was not what this project was about – at which point I deleted the nodes I'd made relating to careers.

Catch-all coding?

Coding strategies involve balancing also between completeness and clutter, between coding the text exhaustively and coding enough just to ensure that each idea canvassed is adequately represented for analysis purposes. This balance will change as the project develops, with coding becoming more strategic. Unless there's a particular reason for picking up every possible mention of something, code only those texts which clearly illustrate what the node is about (the exception here may be coding large chunks for context). Ask before coding: "Why am I doing this?" (Patton, 1990: 122). Additionally, not everything in your data will be equally relevant (though it is hard to know what is or is not at the beginning), and not all coding is aimed at total retrieval; view coding as a means to an end of pattern finding, exploration and reflection.

If you are anxious about missing out on useful data, there are ways of checking the recurrence of early categories, and the earlier occurrence of later categories using tools for searching text (*cf.* Chapter 7). So, move on with confidence, knowing that you will be able to run those checks before you move into a final analysis stage.

You might also expect to engage in periodic re-reading of earlier material, when salience of particular texts becomes more obvious (Miles & Huberman, 1994). In any case, it remains quite likely that you will often create codes which you later drop, and you will occasionally miss relevant passages, but that should not be a major concern. Firstly, important ideas will be repeated throughout the data; secondly, it is likely you will pick up on missed instances of something as you review other nodes, or as you engage in querying your data; and thirdly, you will "wreck your head" (see below) if you try to ensure that you have captured absolutely every relevant passage – quite apart from having far too much coded text to deal with!

In teaching searches (queries) with an earlier version of the Researchers project (used with NVivo 1 and 2), I would demonstrate a way to check whether those who did *not* talk about experiencing intellectual stimulation from engaging in research ever talked about enjoying research. The result was just two passages out of 15 interviews (compared to many which demonstrated a link) – but one of those passages, on re-reading, clearly expressed experience of intellectual stimulation along with enjoyment. Clearly, I had missed coding this passage at that node in my original pass through the document. The appropriate coding (for intellectual stimulation) could be added to the passage while perusing the search results, and (if wanted, although it wasn't necessary in this case as the results node didn't need to be saved) the query could be re-run.

When enough is enough

(*Use of other techniques for checking the rest*)

If coding is becoming routine, with no new categories being developed and no new ideas being generated, it may be time to review your coding strategy. Of course, if your project is one in which it is essential to code all texts in order to thoroughly test hunches/hypotheses or because counting the frequency of occurrence of nodes (e.g., issues raised) is part of the research strategy, you may need to persist until the task is completed. There are some alternative strategies designed to limit your workload which might be used to review additional texts if this is not the case, however. Assuming you have worked with enough texts sufficiently thoroughly to be able to generate convincing answers to your questions or to develop your explanatory theory, then:

- If you have additional data which are not yet transcribed, listen carefully to those to check if any include comments which extend or contradict the model/theory/explanation you have generated from the transcripts you have already analyzed. If you find any, then you will need to work through those interviews (or at least, the relevant parts of them) in detail, otherwise, simply note the confirming remarks.
- If you have additional data transcribed, then import and read through those transcripts, noting any significant instances where concepts you are using or ideas you have developed are discussed. Review these primarily from the point of view of whether they extend or contradict what you have found before. Additionally, you might create a folder for uncoded documents, then use ***text search queries*** for especially relevant terms (*cf.* Chapter 7), limiting the search to the folder with uncoded documents, to check for further instances of key concepts and to review those instances.

There are some, indeed, who would argue that a single case can be a sufficient base from which to generalize for a particular culture, from the point of view that the basic structures of social order pervade the cultural setting, and so will be evident within any case which has been part of it (Barbara Bowers, personal communication; Silverman, 2000).

MANAGING CODING

Managing coding in teams

That different coders will develop and use different codes for the same text sample is an accepted feature of qualitative analysis: what you see in the data will depend to a significant degree on the questions you are asking of it and the approach or perspective you are bringing to it. When the project is a team one, however, it does matter that different coders are approaching the task in a similar way and are being consistent in the way they are using the same categories. This means that there will need to be early agreement on the tree structure of the coding system and the meaning being given to the nodes in it, and regular discussions about how the coding system is developing.

A team working on the basis that the documents to be coded are shared between the members will begin either with everyone coding one or two documents (either the same one(s), or different), or by working together on the first one or two documents before each goes and tries one for themselves. The outcomes of these preliminary coding exercises are discussed, and agreement reached as to what is being sought from the data, and how that might be structured. At this stage it is more important to agree on a broad structure than on particular nodes. If the structure is agreed, then coders will know where to place new nodes, should they need them. These will then become the subject of the next round of meetings when some consolidation of diverse patterns of creating and naming nodes needs to occur.

There are at least two other ways to manage the process of developing and sharing coding in teams. The team leader may develop a coding system which is then shared among team members, particularly where the team members doing the coding are research assistants rather than primary investigators. Alternatively, different team members may become responsible for coding different aspects of the data across all documents, so that it is the documents that are shared, rather than the codes. Whichever way the team decides to work, there will inevitably be a need for regular discussions around the issues being raised by coding, given the inherent flexibility of the task.

Lyn Richards (2005: 100) offers several guidelines for "doing consistency in teams":

- Don't assume inconsistency is bad … It may be the stuff of highly productive debate.
- Ask why a difference in interpretation is there, what it points to, how to work around it, rather than how to eliminate it. That way you may get some deep insights into your data and the progress of your analysis.

- Locate the categories being used differently and make them the agenda for the next meeting.
- ... As a result of the discussion your coding may converge. But it may stay (reliably) different. If so make the categories identifiably different, by name and description. This is not evidence that your analysis is wrong. It is just more data.

NVivo offers a number of tools to assist researchers working in teams:

- The node structure from an established project can be imported into an empty project for new team members, thus providing them with a ready made coding system without the 'clutter' of existing documents.
- A whole project can be imported into another (with the exception of models), so that the work of different team members can be merged before final analyses are run.
- Two copies of the same document, coded by different people using the same coding system (in the same project), can be compared for coding consistency.

Check the options under **File > Import Project** and **Tools > Reports**.

Importing a project will change the target project. You would be well advised to make a backup copy of the original project before importing another project into it.

If a document has been edited in any way that means the number of characters in it has been changed, NVivo will not recognize it as the same document during the merging process. This is particularly significant if you are having different people code different aspects of the same documents, as you will want to bring all those copies together for your final analysis. The best safeguard would be to have team members always work with the coding density bar displayed, as this prevents editing of the document text, or to make the documents read only (via Properties) so the text cannot be changed at any time.

Managing the coding process

When coding was referred to as a 'head-wrecking' activity on the *qual-software* email list, Helen Marshall (2002) analyzed the discussion to identify three ways in which coding could wreck your head. These were:

- through the miserable feelings (frustration, confusion, self-doubt) which can be generated during coding;
- because of the possibility of it going on interminably – there is always something else to be found; a sense that coding needs to continue

until the task is completed, even if it appears that saturation has been reached;

- and because it might impact negatively on analysis – that researchers would focus too much on coding to the exclusion of imaginative, reflective thinking. The problem here was seen as coding rather than computing, however. Marshall quotes the initiator of the discussion as responding to suggestions of computers creating 'coding fetishism' with:

> ... I am a poor veteran of both methods and they both WRECK MY HEAD ... When I used paper and scissors I was constantly chasing scraps of paper – now I am a zombie in front of a confuser. (Marshall, 2002: 62)

From the responses given on the list, she then distilled a number of "ground rules" for managing qualitative coding:

> - Expect that your emotions will be involved, and that some of the emotions will be unpleasant.
> - Respect this emotional involvement. Give yourself room to be reflexive, and plan your processes so that you can periodically step back and view your methods. Ask yourself how you will know when it is time to stop coding.
> - Give yourself time and time out so that your imagination and unconscious can be involved in coding. ...Include the time for "the scholarly walk" in your calculations of expense and progress rates in the planning stage.
> - ...Set routines for coding that will minimise alienation and confusion. These routines should usually involve moving between putting chunks of data in conceptual boxes and thinking about what this means. ...
> - Consider seriously the issue of limiting time spent coding. (Marshall, 2002: 69)

Lyn Richards comments that: "If you've been coding for more than two hours without stopping to write a memo, you've lost the plot." She recommends regularly stepping out of coding, not only to write memos, but also to monitor and review the nodes you have been making. Be free about making new nodes (rather than sweating over each one as you create it), but also periodically do a check to ensure the categories you are making are relevant, and occasionally review one or two in depth for a useful change of perspective on your data.

Additional guidelines for reviewing your coding are in Chapter 7.

NOTE

1 It is, of course, quite possible that a researcher will choose not to use the hierarchical coding structure, but rather choose to keep all nodes as free nodes.

Chapter 6

Managing data

Qualitative data are voluminous and messy! Your data sources are building up and threatening to get out of control. It is time to consider tools for managing them and for storing additional information about them – this is where software excels. Not only will this tidy up your files and help you to see what you have, it will contribute directly to your capacity to answer questions with your data.

In this chapter:

- revisit and extend your ideas about when and how to use folders, sets and cases for managing your data sources;
- discover what kind of information you can store with those sources (as attributes of cases), and different ways of entering this into your project;
- find out how quantitative and qualitative data can be combined for analysis, using attributes;
- use attributes to compare cases and examine patterns in your data;
- filter your data to find a specific subset of cases (or documents) within it;
- preview ways to use folders, sets, attributes and case nodes to refine the process of querying your data.

MANAGING DATA SOURCES

"The most ignored requirement of good qualitative research is that the researcher should be ready [to *handle* what comes] as data" (Richards, 2005: 50). It was the realization that computers could assist with the organization and management of data, rather than a belief that they were appropriate for analysis, that prompted early work in qualitative computing (Kelle, 2004). Over the decades since their early development, computer programs for assisting with qualitative data have become sophisticated toolboxes providing multiple ways to approach the management of data – with your choices about their use being dependent on both data type and methodological orientation.

Managing sources in folders

Just as you group books on the bookshelf, file papers in folders or put nails in a different set of tins from screws, if you have documents which differ in some way it is best to store them in different folders to reflect that. Additionally, by placing documents in folders, you make it easy to query just those sources that are within a particular folder.

NVivo's **Sources** area has three built-in folders for your major types of sources:

- Documents (the data you make),
- Memos (your reflections on the data), and
- Externals (those sources you can't import).

It is up to you whether you decide to create further sub-folders within those folders.

You can place sources in one folder only. Typically the main basis for sorting data documents is according to their type, that is, by whether they are records of interviews, focus groups, literature, observations, or other documents. This is partly to help you keep track of what you have, and also because it allows you to easily select an appropriate subset of documents for particular analyses (e.g., some types of queries that work for interviews do not work with focus groups). Documents from different phases or sites may be sorted in subfolders within those, although, as will be seen below, these are probably more usefully identified using sets rather than folders.

Memos might be of a general type (a general journal, a methods journal) or they may be associated with documents, or with nodes, or just ideas at large (Bringer *et al.*, 2004, provides a useful table summarizing her extensive use of memos in an earlier version of NVivo). If you have multiple memos of several different types, you may find it useful also to sort those into sub-folders under memos. This is either because it makes it easier for you to see what memos you have, or because you might want to restrict a query to one or other type of memo.

If you haven't already created and used folders for documents or memos in the Sources view, you might want to set those up now – simply right-click on the area within Sources where you want to store the new folder, to create it. You can then drag sources from the List View into the folder of your choice.

Managing sources in sets

Earlier you were introduced to sets as a way of bringing nodes together to express a sense that they belonged together conceptually or theoretically. Now you might use sets as an alternative way of organizing sources (documents, externals and memos). You might recall, from your experience with nodes, that a set simply holds aliases, or shortcuts, for the items within it, so that identifying a document as part of a set does not alter that document or the structure of your data sources.

But why bother with sets when you already have your sources in folders?

- Sets are more flexible than folders. Whereas a document can be in one folder only, a document can be included in multiple sets, created for different purposes.
- Sets are treated as a single item when selected for rows or columns of a matrix coding query, whereas selection of a folder of documents results in a separate row or column being created for each of the documents from that folder.

Sets are particularly useful, therefore, as a basis for comparing documents; for example, to compare data from Time 1 interviews with data from Time 2 interviews. Similarly, if you want to compare what your literature says with what you have found in your own data (or perhaps to compare interview data with focus group data), then putting them into sets will allow you to run such a query in a single procedure.

The process of creating sets involving documents is exactly the same as that for creating sets of nodes, or mixed item sets: highlight the required items, right-click, and choose to add to a set or create a set. If the documents you want in a set are already in folders, you can select the folder as a whole and add to or create a set from that. Sets can also be created from documents (or cases) located through Options on the Find toolbar – see below.

Once created, you will find your sets listed by going to **Sets** in the Navigation View. Click on the + next to Sets to see a list, and then on a particular set to see the items in it in List View.

Managing sources in cases

Cases were introduced also in Chapter 3 as a way of establishing units of analysis from your data sources. Because cases are nodes, they can reference (that is, code) parts of documents, one or more whole documents, or combinations of part and whole documents. Thus they add flexibility of a different kind to the way you manage your data sources.

Cases function in much the same way as any other nodes, except that a case always relates to a bounded, definable unit of analysis, rather than a concept. Text or other content can be coded at them, or uncoded from them. Text stored at case nodes can be coded on to other nodes. Cases can be organized on a single level, in groups to identify a range of different case types (e.g., schools, families), or hierarchically to reflect levels of data (e.g., schools, classes, pupils).

The one, critical way in which case nodes are different from other nodes is that ***attributes*** (i.e., demographic or other categorical or quantitative variable-type data) that might become relevant for analysis are attached to cases. This makes them especially important in an NVivo project.

In the majority of projects, any particular segment of data will be coded to one case node only. Where participants are separately interviewed on one occasion only, single documents will be coded to separate cases. If you didn't code documents to cases as they were entered, whole documents can be coded to a case node at any time from the RMB (**RMB > Create Cases**). With multi-person data such as focus groups, you will code each participant's contributions to the discussion to a separate case node for that person. Part documents are coded to case nodes interactively if necessary, or using the autocoding tool if appropriate headings are in place (*cf.* Chapters 3 and 4).

NVivo will allow you to code the same text to more than one case, but this is rarely necessary, and doing so can create problems if you need counts of cases as part of an analysis. The most likely situation where you will code text to more than one case is where you are analyzing cases at multiple levels, for example, for companies and also for units and/or individuals within those companies, or again, for schools, classes and individual pupils. People using these more complex arrangements for cases need to be aware that analyses involving cases will have the same data being counted at multiple levels or places, and to account for that if they are considering counts of cases.

Cases can be used within queries to achieve either within case or cross-case analysis. Any query procedure can be set up to look only at data from a single case, and similarly, you can ask questions that compare cases (*cf.* Chapter 8). Their particular kind of flexibility thus becomes extremely useful.

Using Find to locate project items

Losing track of where you've put something (a document, a note to yourself, a node) can be all too easy! This is especially a problem for academics who dig out their project after a busy marking period (just before conference season) and discover that it takes them some time to get back into it. Find is the tool that will help you locate lost project items – although it does assume you can at least remember what some or all of the name of the item might be!

The Find toolbar is located immediately above List View, regardless of which Navigation View you are in. The kinds of things you can look for are parts or all of names of sources, nodes, sets, models – any item that is in your project. If you're not sure where something is stored, you can search for it through All Folders, and a shortcut to any matching items will appear in List View.

Earlier (Chapter 4) I suggested you could use Grouped Find (under Options in the find bar) to identify all the nodes coding a particular document or case node (or any other node, for that matter). Again, the results of such a request will be shown, temporarily, in List View (make a set of them if you want to store them). Advanced Find (also under Options) can be used to locate those items which meet various single or multiple criteria based on coding, attribute information, or links. Typically this will be used as part of the process of scoping a query, or to select items for a model.

BRINGING DEMOGRAPHIC OR OTHER QUANTIFIED DATA INTO YOUR ANALYSIS

Whether we like it or not, our position in the society in which we live and in relation to the groups within which we move impacts on the way we think and act, and the kinds of experiences we have. At a macro level, an interviewee's sex, class, nationality and religion may singly, or in combination, colour what he or she says in response to an interviewer's questions. Within an organization, it matters what role or position one has, and perhaps how much education or training, or how many years of experience. And at a personal level, attitudes, behaviours, or experiences may relate to one's work history, education, family responsibilities or health. To record these kinds of demographic variables in a project becomes important, therefore, so that their impact can be assessed. In NVivo, the particular values one has on each of these (e.g., Education = tertiary) are referred to as the ***attribute values*** of a case.

Attributes record information known about a case, whether or not it is specifically mentioned in the course of conversation (or other data collection). You might ask people about these aspects of their position in the world as part of your conversation with them; it can be recorded on a check sheet as part of the data collection process; it could have been collected already as part of a larger survey; it might be 'given' by virtue of their location; or perhaps it is contained within archival records.

While attributes are routinely used to record demographic data, you can use them also to record:

- categorized responses to fixed-alternative questions such as are found in surveys; for example, the alternative selected from ☐ often, ☐ sometimes, or ☐ never in response to a structured question about experience of harassment;
- categorical data generated in the course of analyzing the data; for example, whether the interviewee who is caring for mum did or did not mention getting help from other family;
- the circumstances under which the data was collected, such as location, or interviewer;
- scaled responses on instruments designed to measure attitudes or experience, for example, a visual analogue scale measuring level of pain experienced, level of involvement in research, or a score from a standardized inventory; or
- characteristics of a site or organization, where sites or organizations (rather than people) are cases.

Whereas coding is used to capture segments of data which are parts of documents or cases, attributes record items of information which apply to *all* the data for a particular case. Think of it as being a bit like a colour wash – as if all the information stored in a case for a male has been stained with blue, and all that relating to a female with pink. Then anything coded in the text for those cases is automatically stained with their colours. This has several consequences:

- Any further data added to the case will automatically acquire the attributes of the case.
- Attributes cannot be applied to parts of cases. This means that if you want to record some factual item of data for, say, a document within a case (e.g., whether it was the first, second or third interview), this will need to be done by adding it to the document description (via Properties), or if you want to use that information as a basis for analysis, you would need to create a folder and/or set for documents with that characteristic.
- Any coding on the text for a case will necessarily connect (intersect) with the attribute data for that case. This means that text from cases with a particular attribute value that is also coded at a selected node (e.g., females AND time management) can be found using a coding query.

The primary use of attribute data is to make comparisons, for example, to compare the views or experiences of males with those of females; managers with workers; rural participants with urban residents. A secondary use of attribute data is to facilitate filtering of cases, for example, to find (or set up a query to investigate) just those cases resident in a rural area, male scientists under 40 years of age, or perhaps just those documents that relate to the period between 1945 and 1950. For those engaged in a mixed methods project, or who are simply working with responses to open ended questions as part of a larger survey, attributes provide the primary route for combining quantified and free-flowing data (Bazeley, 2006), allowing, for example, connection between interview data and measurements or survey responses (for comparison or expansion), or comparison of examples given following a fixed response question.

- ✔ In order to combine qualitative and quantitative sources, it is essential to have a common participant identifier for both types of data, so the computer can match them.
- ✔ For those with a statistical background, it might help to think of attributes as variable data.

Entering attribute data

Attribute information can be entered into your project at any time. It is stored in a *casebook* – a table in which there is a row for each case, and a column for each

Casebook

	A : Age group	B : Discipline	C : Gender	D : Stage
13 : Cases\group participants	Unassigned	Unassigned	Unassigned	Unassigned
14 : Cases\group participants\Ange	30s	social science	female	experienced
15 : Cases\group participants\David	40s	humanities	male	experienced
16 : Cases\group participants\James	50s	science	male	experienced
17 : Cases\group participants\Jill	30s	social science	female	experienced
18 : Cases\group participants\Marg...	40s	social science	female	early career
19 : Cases\group participants\Paul	50s	social science	male	experienced
20 : Cases\group participants\Sandy	50s	science	female	experienced
21 : Cases\interviewees	Unassigned	Unassigned	Unassigned	Unassigned
22 : Cases\interviewees\Elizabeth	Unassigned	social science	female	early career
23 : Cases\interviewees\Frank	Unassigned	social science	male	experienced
24 : Cases\interviewees\Shane	Not Applicable 30s	humanities	male	experienced
25 : Cases\PhD students	50s 40s	Unassigned	Unassigned	Unassigned

Figure 6.1 Casebook with attribute data

attribute, with values entered in the cells (Figure 6.1). The casebook is accessed via the Tools menu.

When and how you will record attributes in NVivo depends on:

- whether the information is available in check sheets or an existing table or spreadsheet, or has to be inferred or sourced from the documents;
- the volume of data to be recorded; and
- when you need to use it (which is usually not until you have most or all of your data in).

If demographic or other categorical information has to be obtained from the documents, then it is best to set up the attributes you will use ahead of working in the documents. You will then enter the data while you are working through those documents, that is, as you are coding and memoing them.

If the data are available in check sheets or lists of some kind, then you can set up all the attributes and their values and enter the data interactively into the casebook within NVivo either as the cases are created, or for all cases at once.

Alternatively, you can record the data for all cases in a spreadsheet, or you can create a spreadsheet by converting an existing table (e.g., from a statistical database), and that can be imported into NVivo. Importing a table will both create the attributes and enter the values for each case. This is the most efficient method if there is a larger volume of data and/or it is already in spreadsheet or other table-based format. You can import the same table of data multiple times if it is needed for an interim analysis before all text data have been imported, although typically the table data would be imported after all cases have been set up and populated.

If you have rows in your casebook that you think should not be there, you will not be able to delete them directly in the casebook. You will need to delete the case node (but check you don't need the case node for some other reason first!).

CREATING ATTRIBUTES AND ENTERING VALUES WITHIN NVIVO

Create the attributes and their values (Figure 6.2)

- Move to **Classifications** in the Navigation View, and select **Attributes**. Right click in the List View, and select **New Attribute**. Provide a name for the attribute.
- Indicate a **Type** for the values of the attribute. Most attributes will be **String**, as that is any combination of letters and numbers. If the attribute comprises numeric values only, then choose **Number** to ensure that they can be correctly sorted. **Date** values can be entered in local format with day, month and year. If you are wanting to record years only, then use Number, and if you want just month/year use the date format with the same day (e.g., 01) for each month.
- Click on the **Values** tab. Click **Add** and, where you know them, enter the first of the values you will be using for your attribute. Repeat this process for each value that you add. Use **Remove** if you make an error (or later, if you find you don't need this value). You may not know ahead what kinds of values you will encounter for some attributes – simply leave these ones without pre-set values for the time being.
- Repeat the process of making a new attribute and entering values for each attribute you want to record. If you find you want to add another later, that is not a problem – you can come back to this screen at any time.

While age might be recorded as a number, age expressed in ranges (e.g., 0–4, 5–9, 10–14; 20s, 30s, 40s) becomes a string attribute. The same applies to years of service or any similar variable.

The most common error I see when people are creating attributes is that they use what should be a value label to name the attribute as a whole, e.g., they call the attribute Male, instead of Gender. Then they are limited to using yes/no values, and their use of the attribute for comparisons becomes very clumsy.

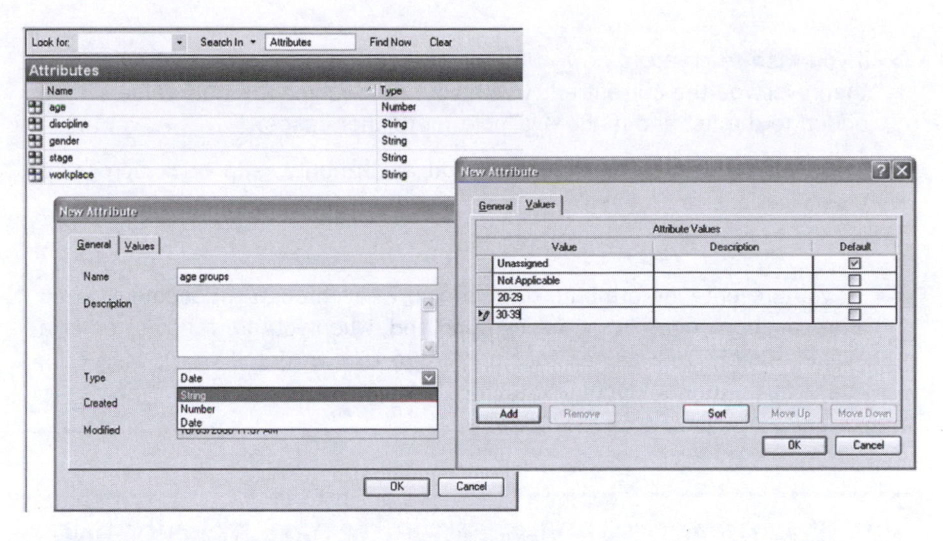

Figure 6.2 Creating an attribute and its value

- ✓ Attributes can take only one value for a particular case. If a case fits two categories (e.g., Fred has two different jobs) then you will have to either: (a) record the most relevant or important value for that person and ignore the secondary one, (b) create a combined category – but only do this if there are likely to be others with the same combination, or (c) create an additional attribute (e.g., Job 2) to record the second value (and again, only do this if there are sufficient with a second job to warrant it).
- ✓ The whole point of attributes is to group cases for comparison, so if everyone, or nearly everyone, has a different value on a particular attribute, then that attribute is not going to be useful for analysis and there's not much point in recording it. This is often an issue, for example, with educational qualifications where everyone lists some particular variant of a degree or diploma – you need to categorize these into a few groups to make them useful. Make decisions like these on the basis of what is likely to be most relevant to your research questions.

Enter values on each attribute for particular cases

- ▶ Open the casebook: **Tools > Casebook > Open Casebook.** The casebook will open in Detail View, with your cases listed as rows, and your attributes as columns (*cf.* Figure 6.1). Currently the values will all be listed as Unassigned (unless you changed the default value).
- ▶ Click in a cell, and select an appropriate value from the drop-down list for that cell (*cf.* Figure 6.1).

- If you wish to create a new value 'on the run', double-click in the cell and then overtype the current entry with your new value. The new value will be added to the list and made available for further cases.
- If you find you need to change or add an attribute, slip back across to **Classifications** to do this.

✓ If you are entering attribute data as you work through a document, keep the casebook open in the background and, when you are ready to enter a value, simply click across to it (using the tabs in detail view), then click back again to your document.

CREATING AND ENTERING ATTRIBUTE DATA BY IMPORTING A TABLE

Create the table

- In a new Excel (or other) spreadsheet, leaving the first cell empty: (a) list your attribute names in the first row of the table; (b) list your case names in the first column, exactly as they appear in NVivo. The first cell (cell A1) will be blank.
- Enter the values for each case throughout the table. Empty cells are best left blank.
- Save the table as **Unicode Text (*.txt)**.[1] Note that text files will not support formatting. Close the file.

✓ The format for case names is best viewed in the casebook itself. Names can be hierarchical, or you can choose to show names only (give all your participants unique identities), by selecting **RMB > Casebook Case Name Format > Name.**

✓ If you have a hierarchical casebook, and you want to import data for all groups at the same time, your spreadsheet will need to use the full hierarchical names (as in NVivo). Otherwise, to use names only, import data for each group separately.

✓ If you are having difficulty getting the names typed correctly, then create at least one attribute (no values needed), then **Tools > Casebook > Export Casebook**; open Excel; select **File > Open**; change the **Files of Type** to **All Files,** select the Casebook and click **Finish** on the Import Wizard. The table will open ready for you to type in additional attributes and values.

- ✓ While SPSS and other statistical packages prefer you to use numeric codes for values (e.g., 1 for male), in NVivo it is better to use text labels (male) as these make more sense when you are reading output from the data.
- ✓ Excel has a habit of converting low number ranges (e.g., 1–3) into dates, so it is safer to write them as, say, 1 to 3. Even if you change the cells to a non-date format in Excel, this information will be lost when you convert and save the file as .txt, so that if you re-open the .txt file in Excel (e.g., to add more data) it will revert to dates.
- ✓ Dates should be entered in accordance with your local settings.

Import the table into NVivo

- ▶ To import the table, go to **Tools > Casebook > Import Casebook**, and work your way carefully through *all* the options (Figure 6.3). If you are not sure about any of these, go to **Help > Importing Cases and Attributes**, expand and check under **import options.** The ones to watch, in particular, are **Case name format,** also the **Options** and **Case Location** at the bottom of the dialogue. A hierarchical table will be imported under Cases. A table with names will be imported under the parent case node for those cases.
- ▶ The Casebook will open (or update, if already open) in Detail View.

- ✓ If you saved your files as Unicode Text, then the default option in NVivo of **Unicode** format should work without any problems. If you saved it as Text

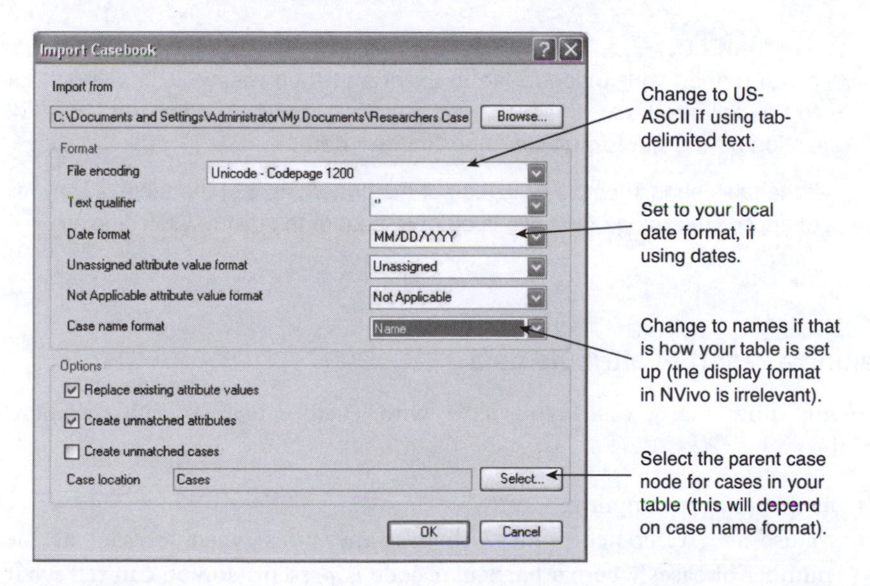

Figure 6.3 Import casebook dialogue

(Tab delimited), then you will need to select **US-ASCII** as the file format in the Import Casebook dialogue.

- If you export directly from SPSS to text format, you will get numeric codes only. To bring data from SPSS, first save it as an **Excel 97 or later** spreadsheet, **write variable names to the spreadsheet** and use **value labels** where defined. Delete unnecessary columns (think about what you will actually find useful in connection with your qualitative data) and then convert it to text format. If you had any empty values, these will become #NULL! in Excel: use Replace (find #NULL! and replace with nothing) to remove them.
- NVivo will automatically recognize the best format for imported attributes, e.g., if all the values in a particular column are numeric, then NVivo will designate that attribute to be a number.
- If your case names include commas or apostrophes, when it is saved as a text file, these names will be enclosed in quotation marks, thus preventing recognition by NVivo. Open the text version of the file in Notepad, and replace " with nothing to remove them.
- If your data fails to import: (a) recheck through all the options; and (b) note, from the casebook, whether any of the new attributes were set up on your previous attempt, and if so, where the import appears to have stopped. This can provide a guide to where a problem may be in your original table.
- Case information in the table can be in a different order from the case names in NVivo.

- If you reopen your Unicode file in Excel and then resave it, it will default to tab-delimited text, in which case you will need to choose the US-ASCII option as the file format for importing the data.
- If the table ends in empty cells (i.e., at the bottom right of the table) it may fail to import. Rearrange the rows in order to ensure the last row is complete.

Making a report of attribute data

Knowing how many cases you have with each value of each attribute is useful:

- as a check on sampling;
- to use as a reference point with base measures when looking at the number of cases where a particular code is present (so you can tell what proportion of a particular group talked about *xyz*).

REPORTING ATTRIBUTE DATA

- Create a summary of attributes, values, and numbers of cases at each value from the **Tools > Reports** menu.

And, if you created the casebook data in NVivo,

- Export the casebook as a tab-separated text file in tabular form either from the menus (**Tools > Casebook > Export Casebook**) or using **RMB > Export Casebook** from an open display.
- Open Excel and **Open** the saved text file (you will need to look for All Files or Text Files to see it). When it comes up with the Import Wizard, just click **Finish**; OR, open a Word document and **Insert > File**, then select the imported data and do a text to table conversion, with tabs (Table menu). Using Excel has the advantage that you can then use pivot tables or graphs to display totals and subtotals of groups in your sample.

Using attributes for comparison

In Chapter 5 you were introduced to the idea of using a coding query to find when the same data was coded for two nodes, such as when a strategy or experience was associated with a particular person or setting or time. The coding query identified all text that was coded with all requested nodes. Thus, I could find all the passages where any participant talked about encouragement or support for being involved with research coming from a supervisor, or, by changing the query, from a friend, colleague or family.

Attributes can be used similarly in a coding query, but to answer a somewhat different question. By using a combination of an attribute value and a node in a coding query, I can find just what was said *by* females rather than *about* females, or *by* scientists, or *by* those over 40 about their experience of encouragement and support – regardless of who provided the encouragement or support. (And, if there is sufficient data to warrant it and the need existed, the analysis could be extended to see which groups talked about which people.)

But attribute values invite comparisons! When I'm thinking about this kind of data, I want to directly *compare* what different demographic groups have said about an experience, an attitude, an issue (or multiples thereof). Is it gendered? Influenced by disciplinary background? Impacted by level or type of experience? How do verbal reports of an experience match up with participants' numeric ratings of that experience? And for convenience, I want to make these comparisons in a single operation. So, although I can incorporate attribute information into a regular coding query, almost always I will use a ***matrix coding query***. This produces a kind of 'qualitative cross-tabulation' in which coding items (usually a

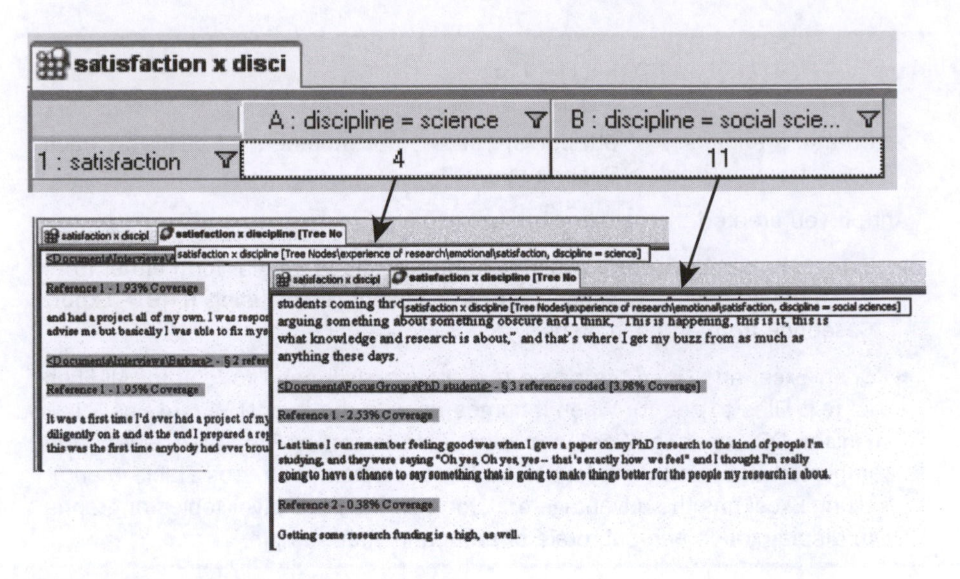

Figure 6.4 Results from a matrix query

node or multiple nodes) define the rows of the resulting table, and the values of an attribute define the columns. Each cell then references text that results from the combination of a particular node with a value of an attribute – a whole series of coding queries put together. These data are initially presented in tabular form with counts of items in each cell, and with each cell able to be opened to show the text that is referenced there (Figure 6.4).

MATRIX CODING QUERY WITH ATTRIBUTE VALUES

- Open **Queries** in the Navigation View, and right-click in the List View, to make a **New Query > Matrix Coding.** Before going any further, check ☑**Add to Project** and provide a name for the query. This will ensure that it is saved for repeated use or modification.
- Open the **Matrix Coding Criteria** tab. For **Rows,** under **Define More Rows > Selected Items > Select > Free/Tree Nodes,** choose one or more nodes to include in your query by placing a check mark against them. After clicking **OK** in the selection dialogue, you will then need to click on **Add to List** in the matrix dialogue. (Rearrange their order using the up and down arrows if you wish, and remove any you didn't intend to select.)
- For **Columns,** under **Define More Columns > Selected Items > Select > Attributes,** check against the attribute *values* you want to include in your query, then **OK** and **Add to List** (Figure 6.5).

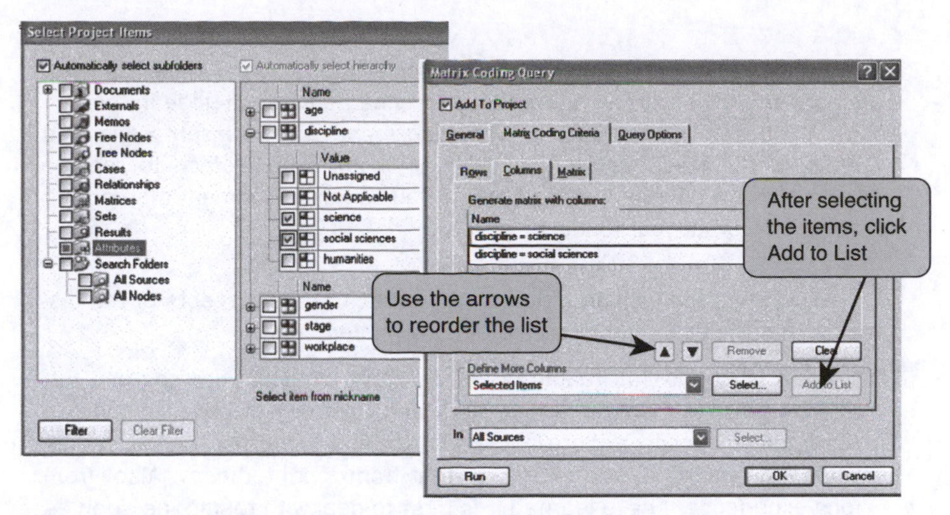

Figure 6.5 Selecting attribute values for a matrix coding query

- The **Matrix** tab is set to **Search for content of rows AND of columns**, which is what you want. This means that for any particular cell, it is looking to find text that matches both the node for that row AND the attribute value for that column.
- Finally, if you want to save the results of your query as a node, you need to turn to **Query Options** and provide a name for the results node.
- To both save and run your query, click on **Run** at the base of the dialogue. Clicking OK will just save the query, ready for you to run when you're ready (right click on the saved query and choose **Run Query**).
- The matrix table will open in the Detail View. Use the icons on the grid toolbar (which will have become active) or options on your RMB to modify the matrix display. Open any cell to see its text by double-clicking (*cf.* Figure 6.4).

Reporting your matrix

- Directly print the matrix table by going to **File > Print.**

  OR

- Export the matrix table to Excel or Word: **RMB > Export Result**, and save in your preferred format (.xls or .txt). If you save as text, open Word and then Open or Insert the saved file, select it, and do a text to table conversion using tabs (Table menu).
- To print out the text in the table: First you will need to Copy or Cut the matrix results node and paste it as a Tree Node. Then, select all the (third level)

child nodes (use Ctrl+Click), and choose to Print or Export. Make sure you select the option to include the **hierarchical name** for each node in the output. If you export, then open a Word document, choose **Insert > File,** and select all the files that were output (at once). Use Replace to replace **Name:** with Name in a heading style so you can use document map or outline view to order your reading.

- If you change the column widths of the table, or other aspects of the display, that will flow through to the directly printed copy.
- My rule is to *not* print reams of results from any analysis program, but to work through them on screen first, and then be selective. In NVivo, I often find I want to explore context or to modify the text or the coding while I'm reading it, and I can't do that from a printed copy. Also, from long experience, I have learned it is best to deal with results as soon as they are generated – and printing them off, apart from using up the world's resources, is a really good way of putting off actually doing anything with them!

On the one hand, the numeric output from a matrix coding query provides a basis for comparative pattern analysis where it can be seen *how often* different groups report particular experiences. Do those with different levels of experience tend to use different strategies? Is one's gender associated with a different pattern of responses to a crisis situation? On the other hand, comparison of the text for a particular node for those in different groups allows you to see *in what way* different groups report particular experiences, and so has the potential to reveal new (or previously unobserved) dimensions in those data. This strengthening of the comparative process is one of the more exciting outcomes of using these techniques (Strauss, 1987).

Researchers' experience of satisfaction in research (gaining personal pleasure from engaging in research) was compared for different genders, and for two major discipline groups. Differences in the way they talked about experiencing satisfaction were not apparent across genders, but comparisons across major discipline areas suggested that scientists and social scientists gained satisfaction from different sources (*cf.* Figure 6.4). Approximately half of the members in each of these two discipline groups expressed satisfaction, but those in the sciences who did so were likely to refer to the sense of agency they experienced in doing research, while most of those in the social sciences made reference to achieving a goal or a task when expressing satisfaction. Although I had noted that a sense of agency was important for some researchers,

I had not associated it particularly with satisfaction (or discipline), nor had I considered satisfaction in terms of its source – more in terms of its consequence. This relatively simple analysis of disciplinary differences in the experience of satisfaction had therefore opened up new insight into possible motivations for researchers, and an additional dimension in satisfaction.

There were, however, three instances where individuals went against the trend: two social scientists (one male, one female) also expressed agency, while one scientist did not. Because the coded text was available, and I was able to immediately access their source documents and other demographic information, these cases could be identified and explored in detail. I found the two social scientists both worked in experimental psychology (which has more in common, perhaps, with science than social science), and the one scientist's recent and current work was all to do with recording the history and biography of science and scientists (which has more in common with social science than science). I could argue, then, that rather than contradicting the observed trend, these apparently discrepant cases added confirmation (or refined the notion of what discipline meant in this instance).

The use of attribute data in a matrix coding query additionally opens up the possibility of corroborating or confirming the meaning of scaled scores by matching scale points with text in which participants describe relevant experience.

Anne Marie Coll, at the University of Glamorgan, Wales, had patients recovering from day surgery use 100 point visual analogue scales to record the level of pain they were experiencing following their surgery. She interviewed them, also, about their experience of surgery and consequent pain. Their descriptions of their experience of pain could be sorted by the rating they had given for the level of pain experienced on the corresponding day (these were collapsed to 10 points for this analysis). In this way, it could be determined what each point on a pain scale of this type meant for people experiencing it, thus making use of this type of scale more meaningful for further research.

For those using survey instruments with both closed and open questions, or following up a survey with open interviews, translating the closed question responses into attributes allows you to analyze the open ended responses, using either whole responses to a particular question or coding derived from those responses, in relation to the choices made on the closed ones (Bazeley, 1999; 2006). Doing so can expand or complement understanding derived from the quantitative responses (Bryman, 2006, Caracelli & Greene, 1997), while consideration of discrepancies between the two forms of response prompts further analysis and (possible) enlightenment (Erzberger & Kelle, 2003).

SCOPING QUERIES

Until now, any queries you might have run have been set to run on all sources. Probably, in the process, you have been making the computer work harder than it needed! Every time you instruct the software to search for a node, or for the association between nodes or between attributes and nodes, the program opens and closes every source – documents, memos, externals, the lot! Saving processor time is just one reason why you might want to limit the items or folders that are checked during the query.

A more substantive reason for limiting how much is checked is when you want to restrict the search to particular sources, node(s) or case(s). In NVivo this is referred to as ***scoping*** a query – a term that suggests one is focusing one's view as through a telescope, to better see one point in the universe. Limiting the scope of the query allows you, similarly, to focus your view on a particular set of documents, or a particular node or case, and then to change that view to explore a contrasting set of documents or another node or case. This facility is available for all types of queries, including simple coding queries, matrix coding queries, and text search queries. In compound queries one can have scoped queries embedded within scoped queries – a level of complexity well beyond most users' needs! I'll keep it relatively simple for now.

Scoping with document folders

Queries are regularly scoped to particular sources, identified by folders or sets. Typically you will have placed different types of sources, such as interviews, field notes, focus groups, literature or other documentary sources, into different folders, so that when you want to limit your query to a particular type of source, you will scope to the folder containing those sources. You might want to find just what the literature says on an issue, or what you have recorded in observations, so you can contrast that with what your interviewees have said.

At a more basic level, whether you have made specific document folders or not, you will frequently want to scope a query to just your documents (and so exclude your memos from the searching process), or to just your memos, when you want to retrieve only your own thoughts on a topic.

SCOPING A QUERY USING FOLDERS

- At the base of any query dialogue is an option to run the query **In** – All Sources, Selected Items, or Items in Selected Folders. Choose **Items in Selected Folders.**
- The **Select** button will become active. Click to choose which folders you want to include in the search process. Selection of a folder will include

sub-folders unless you remove the check mark against **Automatically select subfolders.**

- If you want to scope to just one document, choose **In > Selected Items > Documents > [your document]** (as for Sets, below).

Scoping with sets

For the purpose of querying data, NVivo sees a set as a single item and so the process of choosing to scope a query to a set of documents or a set of nodes involves choosing to run the query *in a selected item*, rather than in items in a selected folder (Figure 6.6). Treating the set as a single entity rather than a group (folder) of items for this purpose will not make any difference to the end result – you will see the finds from your query identified by individual source.

Being able to place sources in multiple sets means that sets provide you with more flexibility than document folders in how you might use them for scoping. And because sets are so easy to make, by selecting folders, items in List View or using Find Options, whenever you want to run a query on a particular group of sources, it is no trouble to 'throw' them into a set.

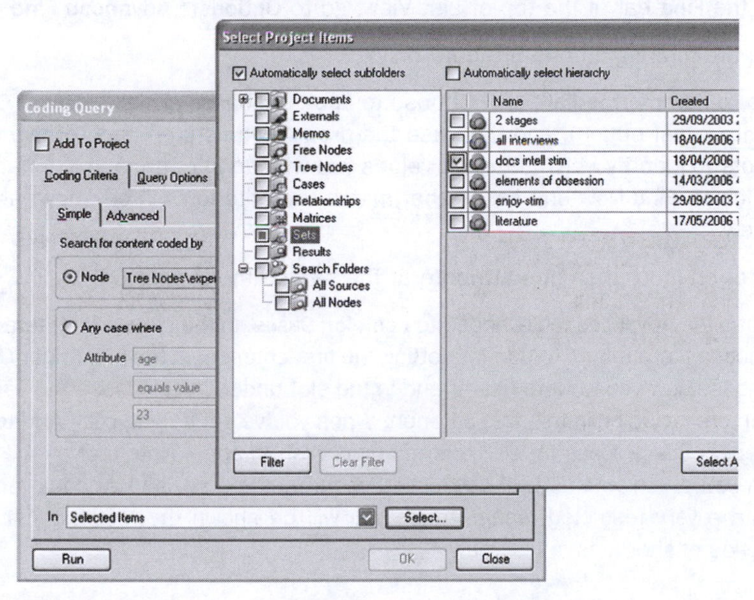

Figure 6.6 Scoping a query using sets

Scoping with attributes

Attributes are used primarily to make comparisons across coded data, but you can use them in other ways as well. Because attribute data are attached to cases, you can use them to identify cases that have particular attributes or combinations of attributes, for example: Do I have any male scientists over 40 in my sample, and if so, who are they? This process is referred to as ***filtering***. You might do this to:

- satisfy curiosity;
- put just these cases into a model;
- make a set of the cases that have some common characteristic of interest; or
- scope a query so that only cases with those particular attributes are considered.

For the first three of these purposes, you will use **Options** in the Find bar, located at the top-right of the List View. To refine (or scope) a query, you can use the results of a find operation by saving it as a set for scoping, or you can use the filter option within the query set-up dialogues. Using find and saving the results as a set is best if you are likely to want to re-use that set of cases as the scope for further queries, while you would filter from within the query dialogue if it was a 'one-off' specification of cases, needed just for this query.

FILTERING CASES BASED ON ATTRIBUTES

- ▶ In the **Find** bar at the top of List View, go to **Options > Advanced Find.**

If you are filtering on one attribute only:

- ▶ Go to the **Intermediate** tab. Choose to **Look for: Cases**. Check against **Cases where** (last option), and then use the drop down lists on each of the three slots to identify which attribute values you wish to filter on, and in what way. Click on **Find Now** and cases that match the criterion will be shown in List View.

To filter on more than one attribute at the same time (Figure 6.7):

- ▶ Go to the **Advanced** tab. Choose to **Look for: Cases.** In the slot under **Interaction,** choose the attribute to use for setting the first criterion; in the slot under **Option** choose how you want to use it; and in the slot under **Value** choose the value/s you are including in this first criterion. When you've got it right, click **Add to List**. Repeat this process for each criterion you wish to apply. Note that all the criteria you put in your list will be applied at the same time, so you could end up with a very restricted result. Again, you will be shown the resulting list as a series of shortcuts in List View.

Filtering within a query

- ▶ Use the same processes as above for Find, but within a query dialogue where there is an option to **Filter**. Once the cases you want have been found (all others will be 'greyed out') you will then need to check the box for Cases to select them as the scope.

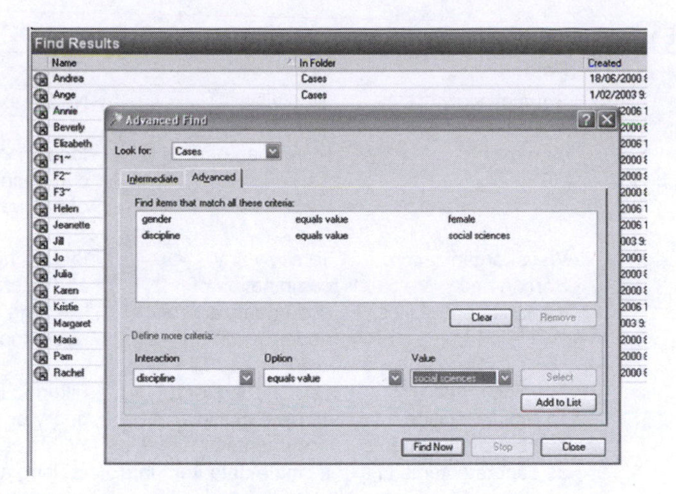

Figure 6.7 Using attributes to filter cases

OVERVIEW

It is not unusual to need the same kind of information stored in a number of different ways within NVivo, so you can use it in different situations. At first this may seem confusing to the novice user, but as you gain experience with the management tools of the software, you will begin to appreciate the flexibility this provides.

Table 6.1 provides a summary overview of these different tools, how they function, when to use them, and how to apply them. An example of how they might be used in a study with complex sources of data is provided also. The application of folders, sets and cases, along with regular nodes, for scoping and filtering queries and models will be explored further in later chapters where you will start to use them regularly as tools for refining analyses.

Example: Managing a complex data set

You are studying the course of events over a period following a traumatic, life changing event for a series of 'cases', with data from multiple time periods and multiple sources. The question is how best to organize and manage the data so that analysis within and across cases can be facilitated.

Your data sources comprise:

- *multiple interviews with each case, obtained at different time periods;*
- *notes from group meetings where multiple cases may have been present, with these notes comprising both your observations of the group as a whole, and contributions by particular cases;*

Table 6.1 Folders, sets or cases: which should it be?

	Folders	Sets	Cases
Accessed in ...	Sources view	Sets view	Nodes view
How many can a single source be in?	One only	More than one	Ideally one only, but can be in more than one (! watch counts)
Primary purpose	Visual organization; Sorting and management of files	To allow any combination of documents and/or nodes	Identifying units of analysis; Locating attribute data about those units
Additional purposes	Scoping a query; To rapidly include a number of documents as separate items in a matrix query	Scoping a query; To treat multiple documents or nodes as a single data item in a matrix query	Filtering in Find, a query or a model; Within case analysis (setting as a scope); Across case analysis (as items in a matrix)
Scoping a query (identifying which data are to be searched)	Can select one or more folders	Can select one or more whole sets (as items within the Sets folder)	Can select one or more cases, folders of cases, or groups of cases based on filtering by attribute
Included in matrix query rows or columns as ...	Member items	Single entity	Individual case nodes; Groups based on attribute values
Special features	RMB to create as set	RMB to create as node	

- *interviews with professionals who deal with individual cases of interest to you;*
- *interviews with caseworkers who are responsible for one or more cases who fall within your sample;*
- *interviews with family carers;*
- *observations of the case interacting with family;*
- *demographic data (age, gender, education, etc.); and*
- *measures of level of functioning at the beginning and the end of the data collection period.*

Management strategies which might be employed for such a situation follow. These are suggestions only, they are not prescriptive, and they will vary depending on the exact nature of the data and the questions to be asked of it.

- *Do not create cases as you import the documents!*
- *Create a document folder for each type of data (individual, group, observations, etc.).*
- *Use sets to identify the source of the data (e.g., professionals, family members, caseworkers).*
- *Create a case node for each of the target cases (the traumatized individual). Code their entire interviews, and those of professionals and family carers associated with them to that case node, also caseworker interviews if the caseworker deals only with that case (this can be done using* RMB > Code > Code Sources at ... *in List View). Data from group meetings or from multi-case interviews with caseworkers will need to be auto coded to relevant case nodes (if headings have been used to identify different speakers within the documents) or interactively coded to relevant case nodes if headings haven't been used consistently.*
- *If each document relates to one time period only, create sets for each time period, and place each document in a relevant time-period set. If documents generally include references to multiple time periods, create a node tree with child nodes for each time period, and code all material (whole or part documents) to a relevant node in that tree.*
- *Create a tree of nodes for various people to whom reference is made within the interviews (e.g., for when a case is talking about the caseworker, or their family).*
- *Import the demographic data and measures of function as attributes of the cases.*

Folders, sets and individual nodes can all be used for scoping a query. Attribute values can be used to filter cases to scope a query to a particular subcategory of cases. Case nodes, tree nodes, sets and attributes can each be used within matrix queries to make comparisons or to identify patterns of association. A combination of matrix query and scoping will allow, say, an analysis of individual or grouped cases over time for a particular factor, or limitation of the analyses to include only the perspective of the cases themselves. The perspective of those involved with a case can be viewed in relation to the characteristics of the case with whom they are involved.

NOTE

1 People working with other languages and scripts should check the various options available to them in Excel and NVivo.

Chapter 7

The 'pit stop'

At various stages throughout your project you will benefit from stepping back from the frenetic pace of moving onward with your data to take stock of where you are at, reflectively reviewing what you have already found and exploring alternative and further possibilities in preparation for moving forward again. By now you will have worked through a number of documents, you have begun to organize your nodes into trees, you are starting to see relationships between different data or coding items. Perhaps this is one of those times for review and revival.

In this chapter, pause to**:**

- View your data from the perspective of the category, rather than through sources;
- Refine the categories you have developed;
- Explore, code or investigate your data using text search;
- Revisit the literature; and
- Play with models, using styles and groups, to build on your case knowledge and refine your theoretical thinking.

SEEING DATA AFRESH IN NODES

It sometimes seems that way, but coding should *not* be an end in itself! Always ask "Why am I doing this?" (Richards & Morse, 2007). See coding as a tool to explore what the data are saying and to identify patterns within them. So, before you move through many documents, and regularly throughout your project, take some time out to review your nodes.

In the node, you see data differently; as noted in Chapter 4, data are recontextualized in terms of the concept rather than the case. In viewing your data through nodes, you will become aware of what is repeated, and therefore more generic, in the experiences, events, and stories that people are telling (Coffey & Atkinson, 1996). You will see data linked by concepts and associations – and you will note discrepancies and misfits. You will find you want to reshape some nodes, to look at others in combination, and to explore responses that don't quite fit.

In NVivo, there are several strategies you might adopt as you review your nodes:

- Gain an overview by generating a summary report of the nodes you have made. This gives a list of nodes with a summary of how many words, paragraphs, passages, sources and cases have been coded at that node.
- Make a list of your nodes. It helps to see, all at once, just the range of concepts and categories you are working with.
- Open up and read the text stored at particular nodes, so you see the data in the context of the category, rather than the original source. Of course, wherever necessary, you can gain immediate access to the original source to see the data in its original context, perhaps to explore why this segment seems to be a little different from others.
- Record what a node is about in a description for that node.
- Record thinking and insights about the concept in a memo for the node. Node memos provide the foundation for later writing.
- Refine the content of a node: uncode what is irrelevant; recode what would fit better elsewhere; code on to new nodes as you rethink and develop your coding.
- Examine what else the node is coded at – and record the hunches that suggest what is related to what in the node memo or project journal; create a 'tentative' relationship node, or check out possible associations with a coding query.

As you move more deeply into your data, don't be surprised if you find yourself spending as much time in your nodes and thinking about nodes as you spend in documents. Indeed, one of the signs of a maturing project is that the ideas being generated from your sources, captured in memos and nodes, become more important than the sources themselves.

Seeing the whole (nodes in overview)

As suggested in Chapter 5, nodes 'tell' a project. A summary overview shows the range of concepts you are working with. Their titles tell if you are working descriptively, or whether you have begun to abstract ideas from the text. An overview indicates which nodes are being used extensively, and which code just one passage in one document.

Does the range of nodes you have created from these early documents reflect the focus and range of questions your project is designed to answer? Are the categories and concepts embedded in your questions represented by nodes, and are these the nodes that are being used more or less extensively for coding? Answering these questions from your summary overview of nodes will assist you to determine whether you need to adjust your sampling strategies, the questions you are asking of your participants (or other data you are generating), or perhaps your strategies in handling the text. Or do you need to modify your research questions? Given, also, that you are looking for recurring regularities in the data, you might take a special look at nodes which code only one or two passages to determine whether they still need a place in your project.

Node lists provide a historical archive of the development of your project. When you make a list of nodes, therefore, store the report (with its date) for review when you are writing your methodology, or to remind you of the development of your ideas as you worked toward your conclusions.

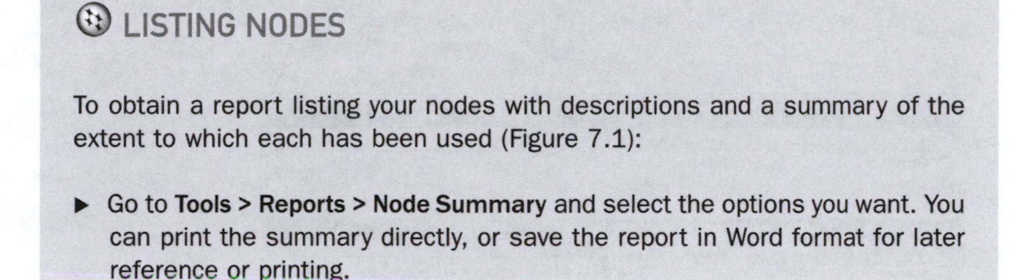

LISTING NODES

To obtain a report listing your nodes with descriptions and a summary of the extent to which each has been used (Figure 7.1):

- Go to **Tools > Reports > Node Summary** and select the options you want. You can print the summary directly, or save the report in Word format for later reference or printing.

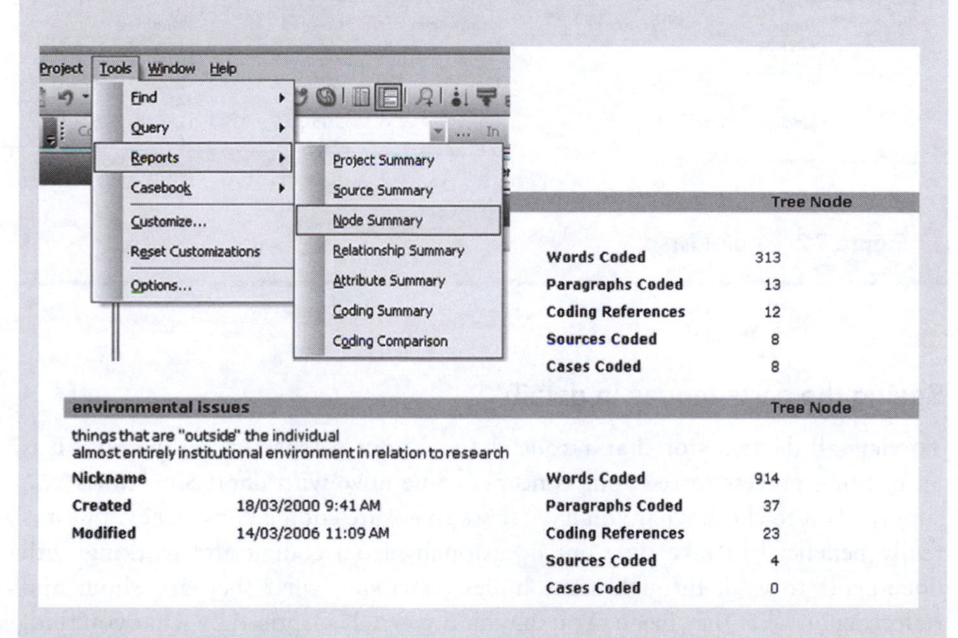

Figure 7.1 Node summary report

To obtain a simple list of nodes (Figure 7.2):

- From the display of nodes in List View, **RMB > Export > Export List**. Select the format you want for your list – so that it will open either in Excel or as a table in Word.

✓ The list may include only the nodes that were on display in the List View, so (for tree nodes) ensure that all those you want in the list are expanded.

OR

✓ Display **All Nodes**, export to Excel, use the node type to sort the display (Data > Sort), optionally delete the Case Nodes from the list, delete unwanted columns, and copy to Word or print from there.

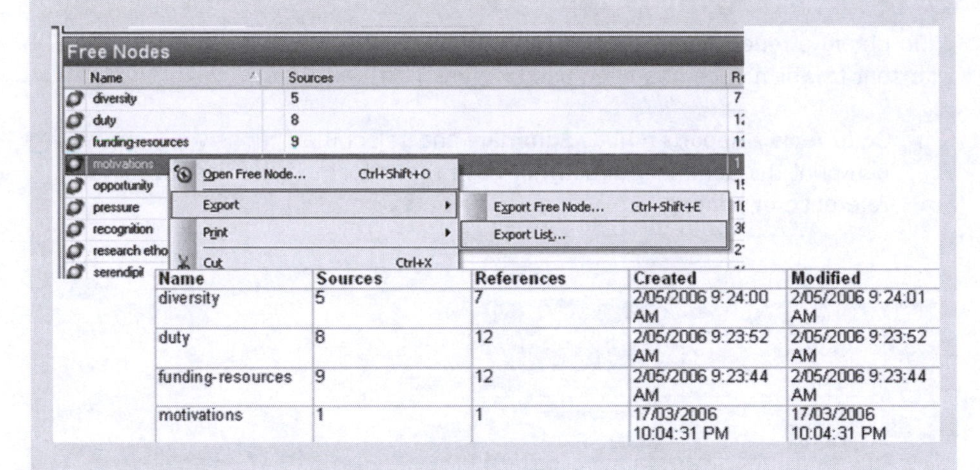

Name	Sources	References	Created	Modified
diversity	5	7	2/05/2006 9:24:00 AM	2/05/2006 9:24:01 AM
duty	8	12	2/05/2006 9:23:52 AM	2/05/2006 9:23:52 AM
funding-resources	9	12	2/05/2006 9:23:44 AM	2/05/2006 9:23:44 AM
motivations	1	1	17/03/2006 10:04:31 PM	17/03/2006 10:04:31 PM

Figure 7.2 Node List

Seeing the parts (nodes in detail)

To view all the text stored at a node changes your perspective on the text. It is an exciting process to see your concepts come alive with data! Sometimes you simply want to check what you have there to ensure coding consistency, but it is really beneficial to take time out occasionally from coding and working with documents to walk through some nodes, clarifying what they are about and reflecting on what they mean. You may find yourself surprised by what you find!

In earlier versions of NUD*IST and NVivo, the term 'browsing' was used to describe the activity of reading the text stored at a node. *To browse* suggests taking a kind of leisurely wander across the pages of text (as through current journals or books in a library; or a cow grazing across a field), picking up points of interest, digressing to examine something fresh, processing and absorbing food for thought, and possibly rejecting some components.

When you view text coded at a node, each passage is referenced so that you know which document it is from. Most importantly, the node gives you 'live' access to the data, which means that while you are browsing, you can access the original context of the coded passage (the paragraph or heading level it came from); jump to the original source in which the retrieved passage will be highlighted; annotate and add links to the text while you are viewing in the node; or add to, delete or recode passages referenced there.

Don't print out all your nodes! Although you are taking time out to go through each of your nodes, it is with a view to refining them, recording ideas, and using them in combination. Printed text is 'dead' data that allows for nothing more than to descriptively summarize. Your purpose is to analyze, involving you perhaps in returning to rethink what someone has said, perhaps in moving forward to develop the concept theoretically as you code into finer dimensions, and most certainly to ask questions about who thought this, when it was evidenced, what conditions encouraged it, and so on. For this you need live access to data at the node.

When you have completed your review, you might record a ***description*** for the node, defining or summarizing it, noting its boundaries – what it includes, what is excluded. Descriptions can be included as part of a printed summary report to create a codebook for your project; this is especially useful in a team project where categories are being shared.

Where too many detailed or descriptive nodes have been created, the exercise of writing a description for each helps you to see repetitions and commonalities among them. Seek to justify each node, as a way of targeting nodes for merging, deleting or 'putting into storage' (create a tree for archived nodes).

If your review of the node gives rise to interesting ideas, or is at all interpretive or abstract, make a linked ***node memo*** to record what made you see this as a category, and write also about what other categories this category might link to. Whereas a document memo stores background information and the ideas and thoughts generated by the particular document or case, a node memo is likely to contain a record of more conceptual thinking, insights from the way the concept is apparent in the text, from making comparisons, possible dimensions, patterns observed. The node memo holds the story of the category, and will become invaluable in the analysis and writing up phases of your project.

For less significant nodes, record those thoughts in the project journal – and remember to date them, reference them (with which source prompted the thought) and code them!

REVIEWING CODING AT A NODE

- Select a node in List View and double-click to open it. Text coded from all sources will be displayed in Detail View.

OR,

Choose to make a Simple Coding Query, and use the tools for scoping the finds to see the text for just some of your documents, e.g., to see just what your memos record about the issue.

- Check the context of a passage: right-click on it, select **RMB > Coding Context** and then **Paragraph** or **Heading Level**; or **RMB > Open Referenced Source.**
- Access the node properties (**Ctrl+Shift+P**) to record a **Description** for the node.
- Create a linked **Memo** for the node (**Ctrl+Shift+K**) to record more reflective or analytic comments. Open the linked memo at any time using **Ctrl+Shift+M** to view or add to those comments.

✓ Accessing the context of the node does not include that context in the node – it is simply a view option. If you want to spread the coding of a passage to the context for that passage, choose **RMB > Spread Coding**, or, select the required additional text while it is in view and choose **RMB > Code Selection at Current Node.**

Reporting a node

- Make a report of the text at a node, or directly print it: Select the node(s) in List View, then **RMB > Export** or **RMB > Print.** Each of these options makes a separate document for each node selected.

✓ Before you print, consider the trees! What benefit will there be in your printing the text, rather than accessing it live? (You can't modify what is there from the printed version.)

✓ Combine multiple node reports into a single document for printing in Word (*cf.* Chapter 6, *Reporting your matrix*).

Checking thoroughness of coding

A combination of text search and coding queries can be used to check for instances where a keyword that has not already been coded to the relevant node appears in the text; for example, you can find all the times the word enjoy appears, where it has not already been coded at the node enjoyment. This will be covered in detail below, as a special application of text search queries.

Modifying what you see

Commonly, as you review nodes, you will want to change some coding. You realize you have more than one concept here; the node is too broad in its coverage; you want to ensure that all the text at this node is also coded for context; these two nodes may really be about the same thing; or you simply find yourself staring at a segment of text, wondering how it got to be there (oops! must have dragged to the wrong spot). Or, on a more conceptual level, as your project has progressed you've rethought what it is you want to achieve, and that means reshaping your data.

If you are unhappy with what you have coded at a node, text passages at nodes can be uncoded (deleted from that node), recoded or coded on to new nodes; and whole nodes can be renamed, 'split', merged together or deleted. This kind of flexibility in handling data and the ideas and knowledge generated from data is essential to an emergent analysis process.

Uncoding

Uncoding removes a text reference from a node. The way you approach this task will depend on the context within which you are doing it, whether that is simply to 'clean up' in a node or to move the passage from one node to another.

UNCODING TEXT FROM A NODE

If you want to move text to another node, involving uncoding and recoding, make sure you do the recoding first – if you uncode first, you have lost immediate access to the passage for recoding!

When selecting a whole passage to uncode, or a part that includes the end of a passage, select through to the right margin, to ensure that all hidden characters (such as paragraph marks) are included.

Uncoding from the current node, using coding bar or keyboard

- Check what node is currently showing in the coding toolbar (unless you have done further coding since opening it, it should be the one you are currently browsing); select the text to be uncoded; then click (Uncode) in the coding toolbar, or **Ctrl+Shift+F4** on your keyboard.

Uncoding from an existing node or multiple nodes, using the RMB or keyboard

- Select the text you need to uncode, then press **Ctrl+Shift+F2** on your keyboard, or hover and right-click to choose **RMB > Uncode > Uncode Selection at Existing Node**. Only those nodes which are coded for that passage will be available to select in the display.

Unless you are looking only at Free Nodes, it might be easier to find the node/s you want by electing to display **All Nodes** rather than Tree Nodes.

Uncoding using the coding stripe display

- If, when you are reviewing a node using coding stripes, you decide that a passage should not have been coded at a node showing in the stripes, right-click on that segment of the stripe and choose **RMB > Uncode**. Just the adjacent passage will be uncoded. Note: this method cannot be used to uncode from the node currently on display, and will not affect that node.

Recoding

If you find it difficult to define the node, and/or it is about more than one thing, or it just happens to contain 'stray' passages, then you might consider recoding text from that node to other nodes. You can code from within a node – the procedure is exactly the same as when you are coding from within a document. When you do so, the coding you do is added to the original document (because all that is stored at the node is a reference to that document, not a copy of the text), so the coding will be present wherever you view that text.

Once the text is recoded, you will want to uncode it from the original node (this, too, will be recorded on the original document).

My review of the node intellectual challenge part way through the project revealed that I had been coding two quite different ideas into that one node. The node included text about the challenge of the research task: "*I find the writing hard, some words don't come easily to me*", as well as about the challenge of solving a research puzzle: "*[the compounds] seemed to be adopting some different shape and because of that, more out of interest, because we couldn't understand it and it just got us sucked in.*" These different ideas needed, then, to be separated, and so I recoded the second group from the challenge node into another existing node for *intellectual stimulation*.

Coding on

Perhaps you are reviewing a broad-based node that can now be coded on in more detail, or into more refined concepts. *Coding on* from already coded text, a term coined by Lyn Richards, is distinguished from recoding, on the basis that it is coding to reflect a conceptual advance, rather than just recoding to tidy up or to better sort the text (Richards, 2005: 97–8). It is a strategy that is used extensively where broad-brush coding has been used for first-pass coding of sources, but it is also used as a regular part of node review. Although the thinking processes might be different from recoding, in practice the same coding process is used to transfer text from this node into other existing or new ones. In this case, however, you may not want to uncode it from the original node.

In a study of children's self management of asthma, Jacqueline Tudball (a PhD student at the University of New South Wales) had a node for physical activity. Her review of this node found that children might use physical activity in order to stay fit and healthy and so to prevent asthma attacks, or they might reduce physical activity as a way of managing asthma attacks when they occurred. This led her to develop a new tree in which she then coded whether strategies adopted by children (physical activity, and also text coded for a range of other strategies) had a preventative or management focus.

RECODING AND CODING ON

- Select a passage of text which needs to be recoded or coded on.
- Display **Coding stripes** > **Show Nodes Most Coding Selection**. The display will show the seven nodes[1] most frequently coding that passage. Use this as a way of checking whether the text is already coded at the alternative node you are considering using.
- If necessary, code selected text to a new or existing node (this may not show immediately in the stripes).
- **Uncode** the selection from any node that this text should not be coded to.

Comparing and combining

After going through the review and description process, you might be wondering why you needed more than one node for that idea. Surely they are about the same thing? Isn't time management all about prioritizing? What's the relationship between ambition and valuing status? And just how does satisfaction differ from enjoyment; or persistence from commitment? What's needed is a way of checking.

The strategy I find most useful here is to make a new node which combines the two (or three) original nodes (Ctrl+Click to select, Copy and Merge Into New

Node), and then display the combined text with coding stripes showing for just the nodes which were copied (Coding Stripes > Show Nodes Coding Item, to select). This allows me to evaluate whether there is a discernable difference running through the text I'm reading, and whether that relates to the way I originally coded the nodes. I can also tell whether there was much overlap in the original coding, and if so, the significance of passages coded at just one, compared to both. If there is significant overlap, I extend my investigation by using a Coding Query (Advanced) to see what is left at one node when I remove what is coded by the other (A, not coded by B). It may be worth while, as an additional strategy, to run a Grouped Find for each of the original nodes (find Items Coding the node) to see if they have similar or different patterns of association with other nodes.

The consequence of such an exercise may be to decide to merge the nodes; to make one subordinate to (a subcategory of) the other; or, to reinforce their difference. Whatever, the outcome, it warrants a note in the description field, and/or an addition to the node memo.

SEARCHING TEXT

When hopeful students ask whether NVivo will find themes and analyze their data for them, it is a text-searching kind of function they are usually thinking about. The capacity to search through sources and identify passages where a particular word, phrase or a set of alternative words are used as a pointer to what is said about a topic unquestionably offers the hint of a 'quick fix' to coding, at least where an appropriate keyword can be identified. The conundrum is that searching text offers much more than that, but also much less. As Richards (2005) notes, searching for words in the text is a mechanical process, so why would one expect it to help in interpretive research?

Text search finds whole words or phrases in the text and, like any other query in NVivo, offers a preview of finds, or codes the resulting finds at a results node – the latter acts as a temporary holding place for finds. In so doing, it locates for you some or all of those passages in the texts which might be relevant to what you are searching for. You can specify how much context from the original document is returned along with the search term, or you can view the context of your finds after the search is complete (Figure 7.3). It is then up to you to review the found passages, and to determine which are actually useful and which are not. Those that you want to retain are best coded on into a new or existing node. You can then delete the original (temporary) results node.

The precision with which text search can be used depends on both the precision with which you can specify the search term and the structure of your data, with the latter impacting on the retrieval of context. If you are planning a study that will rely extensively on text search, then you would do well to experiment and practise with a sample of your data, to ensure that the organization of your

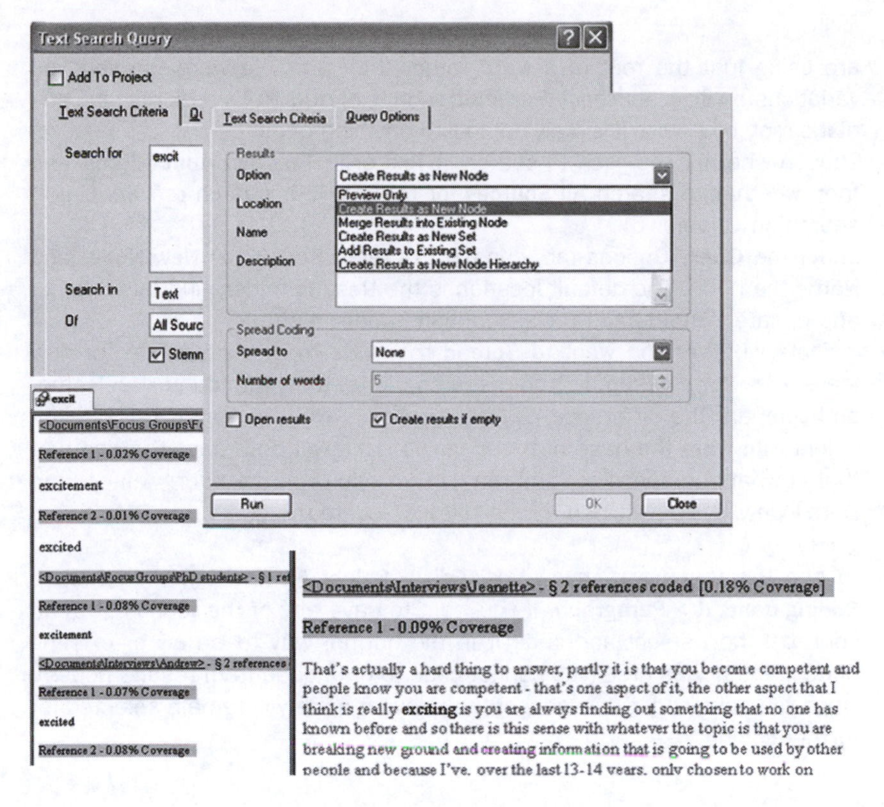

Figure 7.3 A simple text Search

data and the strategies you plan to use will work effectively. Using wildcards with searches or employing text search in combination with other query procedures provides scope for refining or extending the power of text search as a tool for exploration, coding and analysis. These will be considered below.

Try a simple text search or two to see how it works, before reading on to see some of the ways in which you might make use of it, or how you might refine the process.

A SIMPLE TEXT SEARCH

- Navigate to Queries, and with Queries showing in List View, choose to create a **New Query > Text Search. Add to project** (unless you know exactly how to specify what you want to find).
- Type the word or phrase you wish to search for into the dialogue, on the Text Search Criteria tab. This works very much like doing a Google search. If you want an exact phrase, enclose it in double quotation marks. If you

are using just the root of a word rather than a whole word (so you find variations on it), either check Stemmed search, or add an asterisk (*) to the end of the root, otherwise it is likely not to find anything at all.

- Choose whether you wish to search in **Text** or in **Text and Annotations,** and then whether to search **All Sources,** or to limit your search (*cf.* Scoping a search, in Chapter 6).
- Under the **Query Options** tab, choose to **Create Results as New Node,** and **Name** the node. The default location is the **Results** folder, and this is quite appropriate. (The default Preview option is not helpful.)
- Indicate whether you want to **Spread** the finds from the search: for now leave it as **None.** (Then, in true research mode, try it with context selected, and compare the difference!)
- Select **Run** from the base of the dialogue. This will both save the query (if that is what you specified) and run it. The results of the query will open in Detail View. (If you accidentally hit **OK,** just locate the query and **RMB > Run Query.**)
- To see the found text in context, **Edit > Select All** (**Ctrl+A**), then **RMB > Coding Context > Paragraph.** If you want to save any of the found text (and context), then select and code it in the normal way to a new or existing node in the nodes area (you cannot code to or uncode from results nodes). Once you have coded a passage, the relevant node will remain selected for further coding.

✓ Text search will not find part words, symbols or punctuation, or 'stop words' like 'a' or 'the' (see linguistic analysis, below, for a further note on the latter).

✓ The root of a word for stemming is all but the suffix – text search will not find anything if you put in less than that. If a stemmed search isn't finding what you would expect, then use a search for alternate words, or add an asterisk to substitute for any additional characters (*cf.* **Special** on the text search dialogue).

✓ Stemming does not work for languages other than English (use an asterisk, instead).

✓ If you have a lot of finds, don't stress the computer by trying to view the context of all of them at once!

✓ Always treat the node with the results of a text search as a temporary holding area only; once you have coded relevant text on to a more permanent location, delete the temporary node.

! Never send the results of a text search into an existing node unless you know exactly what the search is going to find, and you are sure you want all of those finds in that node.

Do a stemmed search for *excit* in all Documents, and you will find exciting, excited, and excitement. Select all the finds and expand to see the surrounding paragraph. This search illustrates a number of the issues that arise in searching text.

- Some of these finds are located in the same paragraph (e.g., for Frank, Jeanette), which means that using the method of asking for the context *after* the search is completed produces a separate copy of the whole paragraph for each find within it.
- When you have multiple finds in the one passage, you then need to determine whether the adjacent finds should be coded in a single passage, or whether there are different ideas being expressed, so that you would code these as separate passages.
- Further checking will reveal that the find for Barbara is not relevant to being a researcher, and so should be ignored. One might argue that the first find for Andrew is also not relevant, although that position changes when the second find is considered.
- Even with the paragraph retrieved, the first find from Focus 4 is not interpretable. Use **RMB > Open Referenced Source** to identify sufficient context to decide relevance (and then code the preceding paragraph as well, so that the context is retained).

Some of these issues, especially the first two, are more easily resolved if paragraph context is asked for from the start. This creates an alternative issue, that it is not so easy to identify the target word within the paragraph, especially if some of the paragraphs are long (as is that from Andrew, for example). The solution to this, if just one word is being searched for, is to use **Edit > Find [search string]** to locate the word within the paragraphs.

Despite these various issues, the search for *excit* did achieve its goal of finding some (if not all) of the ways that researchers talked about the excitement that comes through research.

What can you do with text search?

Now you have seen what happens when you ask NVivo to search through your text for a word or phrase, let's look at the multiple ways you might use this versatile tool. Here are some of them:

Exploring data with text search

Text search is perhaps best seen as a tool with which you might explore data – and there could be many reasons why you might want to go exploring.

- At an early stage in your coding, check whether something being mentioned is likely to be a topic that will get much attention. Frank, for example, starts by describing the role his PhD played in his development – do other experienced researchers also refer to their PhD experience? Is this an area that warrants particular attention when seeking or reviewing data from new participants?
- At a later stage, if you become aware of a new issue or topic or your thinking about the project takes a new direction, quickly explore across your data to see if this is a viable area for deeper investigation.
- Similarly, in the course of working through a later document you might come upon a new concept or category that is 'nodeworthy',[2] and then wonder whether it was there before, but you just didn't see it. Use text search to rapidly check whether, indeed, such things were mentioned earlier (such a search might be scoped to just the already coded documents).

Such searches depend, of course, on whether you can identify suitable keywords to 'get a handle' on those concepts or categories – an issue which becomes even more pointed when you want to use text search as a tool for coding.

Coding using text search

Text search works very effectively as a way of locating all the paragraphs in which a person is mentioned, a topic raised or a question asked, as a starting point for review of that topic. With your data sorted into a broad topic area, you can then review the resulting node and check the content and context of those passages, revisit the source document if necessary, determine just exactly what is relevant, and uncode, recode or code on accordingly. This may also be an appropriate strategy for some forms of conversation or discourse analysis, where text search can be used to locate passages for detailed coding and analysis from within the larger body of text.

Searching for sleeping and feeding quickly identified passages (paragraphs) relevant to those tasks in infant care in the first stage of a study of the continuity of micro-cultural care behaviours between home and child care for culturally diverse families, by Katey De Gioia (2003). Analysis of those passages then revealed that this was not so much an issue for the parents, but that what was more important was the quality of communication between the child care centre staff and the families. This new direction then became the focus of further interviewing and analysis.

If you are using NVivo to analyze field notes, whether written in Word or NVivo, you might facilitate rapid coding of those notes by strategically placing routine keywords within the notes you write (or dictate). This would be particularly useful if you are in a situation where you are unable to code as you go.

As a tool for directly facilitating more detailed coding of unstructured text, text search can be less adequate, generating what I typically call 'quick and dirty' coding suitable primarily as a stop-gap in emergency situations. Attempting to use it for more than that, and particularly as a primary tool for interpretive coding, is bound to disappoint.

Some years ago when I was leading a team under pressure to complete a report for a very large project in which some of the interview data had not been coded (Bazeley et al., 1996), we needed to write about mentoring of early career researchers, and we knew it had been specifically asked about in the later interviews. A search for the word 'mentor' pointed to all those passages where the question had been asked or which otherwise dealt with this topic, and so it was easy then to review the original sources, to find what was said and to confirm and elaborate what we had already sensed were the issues around mentoring. We absolutely could not do the same thing with respect to the importance of building a 'niche' area in research – there were no keywords which would unlock that aspect of the texts for us. Similarly, when I try to run a search for time management, or even just 'time' in data from researchers (and probably even from managers), I will find everything I didn't want to know about next time, first time, full time, at the time, whole time – and virtually nothing at all relevant to managing time.

Investigating linguistic expressions

Patton (2002: 454) suggests making an inventory of and defining key terms and phrases used by the participants in a study, as a way of helping the investigator to understand their worldview. Use text search to check if a candidate term (including an *in vivo* code) is, in fact, used widely, by a minority only or perhaps by just one person. The context of its use can be investigated at the same time. Similarly, employ text search whenever you meet a metaphor, idiomatic expression or other linguistic feature of interest, to investigate when and how it is used.

In an Australian study of the health impacts of service on a now decommissioned class of submarines, the topic of jollies *was a 'hot' one for the submariners. An exploratory search of text from interviews with former submariners revealed that they used 'number of jollies' as a somewhat unconventional measurement to estimate length of time spent at sea, and number of ports visited.* "At sea, we spent days dived 130, days surfaced 129, ports visited including Australian visits during 2½ years were 11 ports. That's an average of 4.4 jollies a year." *Interestingly, their estimates differed from the official naval records, which the men claimed were inaccurate or incomplete.*

Perhaps you can identify, also, patterns of discourse among your participants. Are the identified expressions used widely, or are they linked to a particular segment of the population, or a particular context? Do they identify the network or culture a participant moves within? For example, do those in the service professions use words which suggest one-way or two-way exchanges when talking about their work and the people they serve? Do the words used in an interview or narrative tell you something about the mental state of the speaker, or their attitude to the topic of conversation?

When people recover from trauma, as well as becoming better able to report the emotions they felt during the crisis period, they also change the way in which they talk about the damaged part of their body (Morse & Mitcham, 1998). They shift from using an objectified reference ('the hand') to using a personalized reference ('my hand'), while references to that part before the trauma are also personalized. Thus, the use of personal pronouns can be seen to chart the progress of injury and recovery. Applying this concept in a different setting, one of my workshop participants reviewed the use of the, my *and* our *in relation to the way primary school principals referred to their staff. Each form of expression communicates a very different attitude to leadership within the school.*

NVivo Help > Text Search: Special Characters and Operators lists a series of 'stop words' (and special characters) that will not be found even if combined with other words (including *a* and *the*, *is* and *was*, and *not*). Where you are using one of these words in combination with an OK word (e.g., "the staff") the best strategy is to search for the OK word, spread finds to 1 word either side, and then use **Edit > Find** within the results node to locate instances of the stop word and so to code the finds you want to count or keep.

Checking conclusions using text search

In the next chapter you will explore ways of drawing and verifying conclusions from your project. How might text search contribute to these?

- Use text search to test the "dominant theme" you are seeing in your data:

  > Every researcher has a war story of a dominant theme that grabbed their attention, a word that seemed to occur everywhere, but which in the hard light of day was being contributed seldom and then mainly by the members of the research team! (Richards, 2005: 156).

- In projects where more data has been gathered (and imported) than is needed to reach theoretical saturation (often a situation where statisticians have dictated the sample size!), use text search to check through remaining documents to locate and explore any passages that confirm or are at odds with provisional findings.

Refining and extending text search

Searches can be refined to allow for misspelling or typographical errors, or to find alternative words or combinations of words. They can be extended by combining text search (or the results from a text search) with another search or coding query.

Check **Help > Text Search: Special Characters and Operators** to find a detailed explanation of various wildcards and other ways of making your text searching more (or less) specific with:

- wildcards to replace one or more than one character in words;
- Boolean terms (AND, OR, NOT) to specify particular combinations of words; or
- fuzzy (vaguely approximate) or near operators.

The proximity operator (NEAR, e.g., ~10) requires that the two words are not stemmed and are enclosed in double quotation marks. If this isn't finding what you think is there, use a compound query instead.

Combine text search with other searches, either to make your search more focused, or to ask questions involving the use of language. The ***Compound Query*** tool provides for the combination of two separate text searches or of a text search with a coding query, using a NEAR operator to link the two separately specified queries, within a specified distance. This might be used, for example to find:

- two words or phrases which occur in the same paragraph (or other distance) – thus using the more natural context of the paragraph rather than the context provided by a rigid word count (Figure 7.4);
- a text pattern (word or phrase) in the context of the results of a coding query.

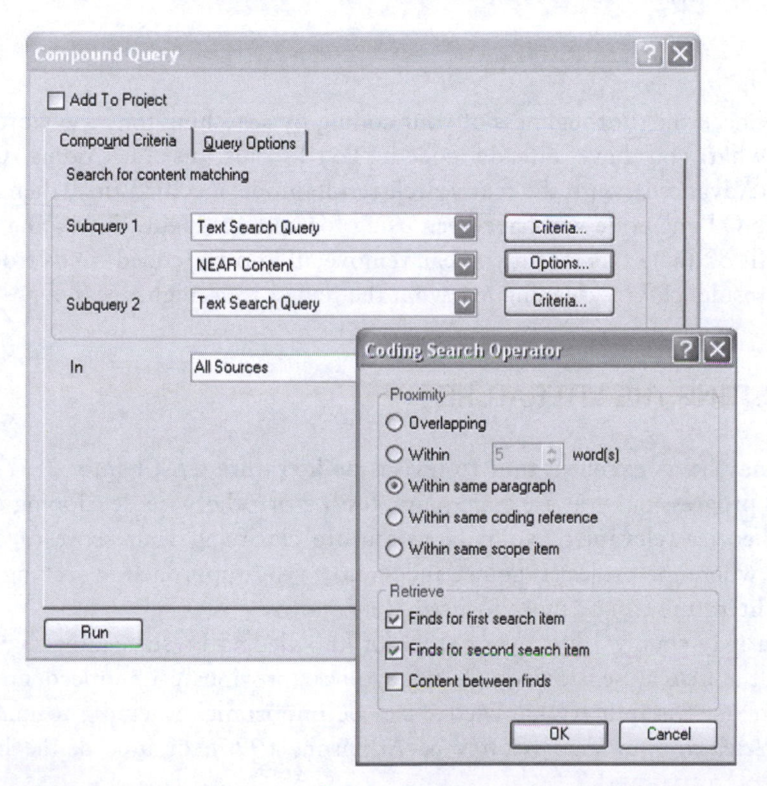

Figure 7.4 Using compound query for text search

Use the results from a text search in another type of query by using the node containing the results as an item in the query.

- Compare usage of words in different contexts or by different groups by using one or more text search results nodes in the **Rows** of a **Matrix Query**, with either contextual nodes or attribute values in the **Columns.**

A matrix coding query was used to compare terms used by researchers to express the idea of being innovative in research, depending on whether they were describing a researcher with ability or a researcher who demonstrated quality (Bazeley, 2003). Not only were words to express innovation used more frequently in regard to a researcher with ability, but the choice of words differed in the two contexts.

	Context: a researcher with	
Word used	**Ability**	**Quality**
creative	29	9
imaginative	11	1
lateral	8	0
innovative	14	15
novel	2	3

- Check the thoroughness of your coding by searching for a keyword, less what you have already coded. To do this, use a **Coding Query** (Advanced), with the text search results node specified first, then AND NOT the node you have been using for coding (Figure 7.5).[3] The benefit of doing this is that you can remove all known (coded) finds from the results before 'cleaning up' what the text search might locate.

REVISITING THE LITERATURE

This may be an excellent time to revisit the literature (*cf.* Chapter 3). Your project is progressing, you are 'emerging', testing or otherwise developing concepts and theories relevant to your data, you are clarifying your sense of direction about where the project is going, and so now, your approach to seeking out and using literature can be more focused.

You may simply add to your stock of knowledge. Fresh reading may refresh your 'theoretical sensitivity', alerting you to previously unnoticed nuances in your texts. Disciplinary knowledge can be important in helping to make sense of puzzles in your data (Coffey & Atkinson, 1996), or indeed, the historical

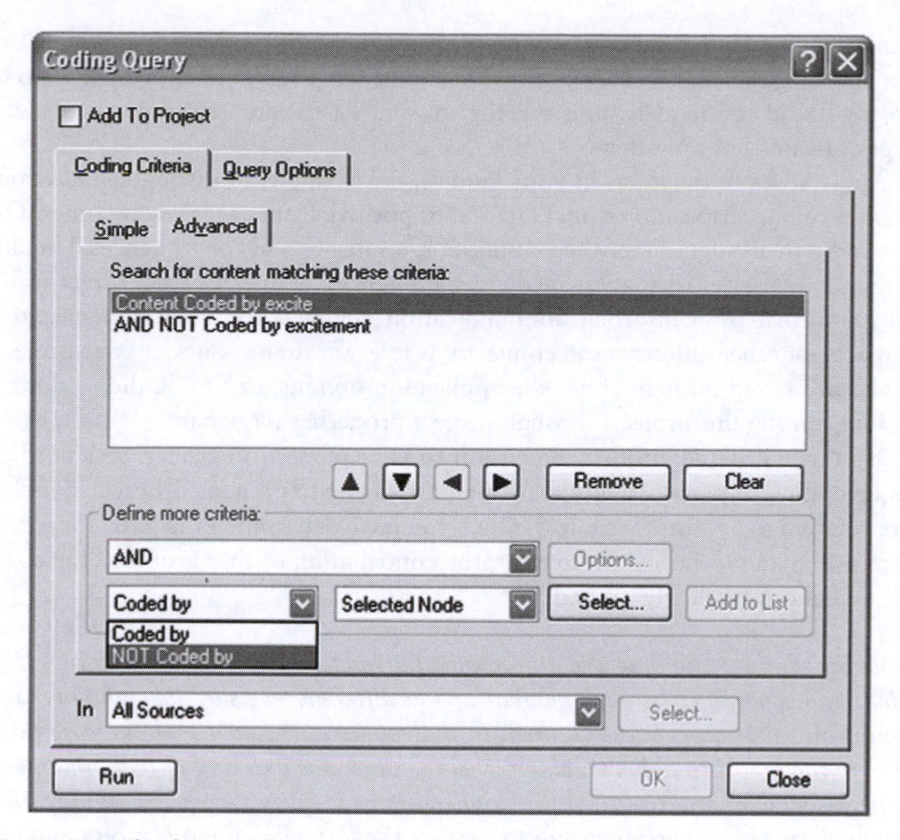

Figure 7.5 Checking coding using an advanced coding query

discourses contained in your literature may provide the foundation for understanding the discourses of your interview texts (Grant, 2005; Kendall & Wickham, 2004).

PAUSING TO 'PLAY' WITH MODELS

Pausing to 'play' with models will also prepare you for the next phase of investigation as you seek fresh data, and progress toward clarifying and testing the ideas you are developing.

Building on cases

If you are following the suggestion to use the modeler to visualize each case as you complete coding of it, by now you should be building up quite a series of models. As part of this review period, evaluate those models, identifying commonalities and differences across the cases. Look for patterns in how you are constructing the models; check out which nodes keep reoccurring, which seem to be pivotal in the models, and whether any are associated strongly with some types of cases but not others.

Apply styles to the items in your models, for example, to distinguish environmental conditions from personal factors, or positive from negative outcomes. Or, use styles to distinguish between conditions, strategies and consequences. Use different arrow styles to suggest mediating influences as distinct from direct influences, formal from informal communication channels, positive from negative impacts, or other differences in connections relevant to the issues of your project. Styles are set up through either the application options, or (if you didn't do this before starting this project) through project properties (*cf.* Chapter 4).

Now you can add another dimension to your models. Items can be identified as belonging to particular ***groups*** which are devised and named by you (and so are referred to as custom groups). Once you have determined the basis for a custom group and identified its content, the contribution of that group to the overall model can be clearly viewed.

Imagine you are studying the communication patterns within a company. You identify a series of cases – people working at different levels in the company, and you draw connectors between them to indicate who speaks to whom. Use position in the model to indicate which level the person is working at (e.g., manager, supervisor, worker). You then use connectors to identify formal communication channels and use a custom group to capture the formal communication network. This is repeated for the informal communication network, as a second custom group. Perhaps some of those in these networks wield particular influence – they are the powerbrokers – identify them with a distinctive style. When you view the model for a custom group, you can see where the powerbrokers lie in the networks, and compare which they are most active in. Once project groups are activated (see below) you can also compare the formal and informal networks (and powerbrokers) in relation to demographic and other attribute values.

CREATING CUSTOM GROUPS IN A MODEL

- At the right of the model window, under **Custom Groups**, select **RMB > New Group**. Name the group, and optionally provide a description.
- To populate the group, select an item or multiple items in your model, then click in the check box under ✔. The same item can belong to multiple groups. Remove an item by selecting it then clicking to remove the check mark.
- To show or hide a group of items in the model, click on the check box under 👁.

 Custom groups are model specific. If you make a new model, even if it is in the same project, you need to re-create the groups. Styles, in contrast, are available to apply to items or links in all models in a project.

Check **Help > model groups**, for further assistance.

When you include multiple case nodes in your model (e.g., in a model of communication networks within a site), then ***Project Groups*** will be activated automatically. These groups are based on your attribute values, and they are used to show or hide the presence of cases with particular attributes within your models. Similarly, if your model contains items linked by relationship nodes, then the Project Groups area will contain a list of relevant relationship types to allow you to show or hide related items of various types.

Models are more often built around tree or free nodes than case nodes (except perhaps where interpersonal relationships are being modeled), and so you may not have many occasions where project groups will prove useful. One possibility is suggested by Miles and Huberman (1994: 199) in their discussions about displaying data. They 'scaled' their cases on each of two dimensions of interest, and then plotted their position against two orthogonal axes, to see in which quadrants and how closely the various cases might cluster. In this scenario, the project groups function could then be used to check the relationship between clusters and various demographic or other quantitative variables.

Building concepts and theory

If you used the modeler to record theoretical connections between nodes earlier (*cf.* Chapter 5), you might now reconsider and refine or extend those models based on insights from reviewing your nodes, and/or the patterns of association you are seeing across your cases. Preserve the originals as static models (**RMB > Create as Static Model**) and then adapt the model to reflect your developing understanding. Use styles and groups to enhance your models.

THE PERIODIC PAUSE

This 'time out' for review and reflection should not be a one-off event! As you move through working with more data – coding, memoing, linking – make a habit of periodically pausing to visualize, to 'hypothesize', to journal as a useful prelude to the next step.

And now you might pause also to sigh with relief and expectation, for you have now met (at least briefly) the full range of tools NVivo offers, and what remains for me is to demonstrate how you might use these various tools to help answer questions from, within and about your data.

NOTES

1 Seven is the default number of stripes. This can be changed through the coding stripes dialogue (bottom option), but if you are asking for nodes coding selected text, 7 is likely to be all of them.
2 An expression often used by Lyn Richards.
3 If, by the time you are reading this, there is a NOT operator in the compound query options, then use that tool to undertake this task more directly, by avoiding the preliminary step of making a node of the text search results.

Chapter 8

Going further

Towards the end of their very useful book, Coffey and Atkinson wrote:

> We have not advocated one particular type of analysis, one particular theory or way of writing. Principled choices of method reflect academic disciplines, theoretical perspectives, foreshadowed research problems, and individual intellectual styles. We would not wish to see all the possibilities of qualitative research being limited by the adoption of common paradigms. We are prescriptive, however, in another sense. We believe there are some basic principles to be adhered to, whatever particular method is adopted. (Coffey & Atkinson, 1996: 189)

It's not possible to provide a definitive ending chapter to this book. This stage of work depends so much on your research purposes and methodological preferences, and your comfort in manipulating a computer. What I can provide is ideas for ways you can take your analysis of data further than 'just code and retrieve' in the context of those "basic principles" of continuous analysis, creativity tempered by rigour and care, thorough documentation, and flexibility along with disciplinary awareness.

Most of the tools you will use to go further have already been introduced to you, and indeed, much of what you have been doing as you have worked through the last two chapters has been progressing your analysis beyond 'just coding'. While I can't give you a recipe for a particular goal or set of data, what I can do here is focus on strategies for using the available tools creatively and systematically as you work through and with your data. In the process, you will gather understanding and evidence to enlighten and convince the world as to the value and veracity of what you are finding.

In this chapter:

- think about what you are aiming for in your project;
- see how to manage queries in NVivo;

- explore ways of using cases in analysis;
- develop core concepts and themes for your project;
- reflect on narrative and discourse features of your data;
- try different strategies for 'emerging' theory, and think about how you might make causal inferences;
- find ways of testing and arguing the veracity of your conclusions.

THE ANALYTIC JOURNEY

Like almost everybody who writes about qualitative analysis, throughout this book I have emphasized that analysis is ongoing throughout your project. You begin with reflective memoing. Coding involves preliminary analysis. Query tools were introduced early so you could start early to ask questions of, and check associations in your data. Theoretical sampling, assumed in many qualitative approaches, involves deliberately selecting further participants (and the data you seek from them) based on their ability to elaborate on issues revealed in the data so far. "In many studies, there are no clear stages: Issue development continues to the end of the study; write-up begins with preliminary observations" (Stake, 2000: 445). Documentation of reflections and decisions on the way is part of the transformation from personal experience and intuition to knowledge. Your analytic thinking may shift in direction during the course of your project, with the final focus being quite remote from the initial problem. Alternatively, it may simply become more refined and focused as your understanding of the research problem deepens.

There is, nevertheless, a danger of becoming 'becalmed' when you have worked right through your documents and you know your data well (Richards, 2005) – where you will stop and wonder, "What do I do now?" Shifting gear from coding to ways of searching and seeing your data is easier if you have been doing it all along. Coding is not an end in itself; it makes sense only if you can use it to search and test the ideas that have been coming out of your data.

Start small, with a question, a concept, a puzzle, and explore from there. Write as you go, and small questions will build into bigger ones. Keep asking "I wonder if ... ?" and explore, test, check back into your data and into your disciplinary literature, building up until you reach an integrated picture. The secret to analysis is in asking questions of the data, and then in thinking through how you might pursue ways of answering them from the data. If you've stalled in the approach you're currently using, go back to the methodological literature seeking fresh stimulation. Read other studies using the same methodology, even if they are on an entirely different topic, for further ideas on methods. Read theory, and wonder how it might inform your study. Check out the last few chapters of "*Handling qualitative data*" where Lyn Richards (2005) provides a wealth of practical strategies for searching, querying and seeing your data as you seek to move your project forward and bring it to a conclusion. Miles and Huberman (1994) also provide a wealth of ideas about ways of displaying your data to help you see what's there, to see the patterns and draw conclusions.

While there is no need to feel you have to follow one methodological approach obsessively – perhaps there are other strategies you can borrow that will help you through an impasse – you do need, at some stage, to become obsessive about your project and your data. Let the ideas you are playing with and the data you have permeate your whole being. Then, when you take time out to walk, soak under a shower, sit by the fire, gaze at the stars or smell the roses, that tranquil activity will allow your brain to process the information that you've been absorbing. If in these moments fresh insights do come (to your prepared mind), write them down (they can be perilously fragile). Think back to what it was you were thinking about; try to identify what led you there – and write that too. Then, when you return to the computer, test the idea against your data and build the chain of evidence. And keep writing!

Where are you going?

"The ultimate excitement and terror of a qualitative project is that you can't know at the start where it will end" (Richards, 2005: 125). By now, however, you should have a strong sense of where you are heading, what you are expecting to achieve, what questions you will answer with your project, what will bring it all together.

Outcomes of your work with your project might include:

- ***'Thick description'*** – a term popularized from the ethnographic writing of Geertz (1973) to convey deep understanding of a culture or experience. Thick description goes beyond details of spoken content to include wider semiotic analysis, attention to context, and other products of careful observation. Neither a 'patchwork quilt' of quotes, nor an 'illuminated description' provides an adequate substitute (Richards, 2005). Similarly, goals of 'letting people speak' or 'giving voice to the participant', of themselves, do not make for analysis but are the subject of more literary forms of writing.
- ***Emergent theory***: Qualitative analysis is typically described as being an inductive process, designed to generate or extend theoretical understanding of the issue or experience being investigated. Your task, as a theory-builder, is to work at identifying and making sense of the patterns and relationships in your data. Think of building theory as just that, especially if your reading of established theoretical perspectives leaves you feeling daunted by the idea that you can do that. Theory is often small and local to start with; as your work extends and grows, so will your theoretical sophistication grow. It needs to be worked at, however, because theory will not emerge on its own.

- *Theory testing* – less frequent as a primary goal for a qualitative project and perhaps more commonly found in projects involving either mixed forms of data or mixed (statistical and textual) analyses. NVivo's query tools do allow, however, for extensive testing of hunches or hypotheses which have been brought to the project or developed within it.
- *Practical application* – in policy analysis, situation analysis, needs analysis or program evaluation.

> Not all questions are theory based. Indeed, the quite concrete and practical questions of people working to make the world a better place (and wondering if what they're doing is working) can be addressed ... there is a very practical side to qualitative methods that simply involves asking open-ended questions of people and observing matters of interesting real-world settings in order to solve problems, improve programs, or develop policies. (Patton, 2002: 135–6)

The question now is: How can NVivo assist you in moving forward?

QUERIES IN NVIVO

Searching and asking questions of your data in NVivo is managed primarily through the Query tools, most of which you will have seen already. Particular strategies for analyzing data and generating results will be discussed further below, but first, a reminder of the general principles by which queries in NVivo operate.

What do queries do?

When you run a query, NVivo locates all the passages that meet the criteria you have set in your query. Depending on how you have set your preferences, additional information typically provided with the actual data identifies the source it has come from, and also gives an indication of how many finds there are in each source, and how much of each source was found. Your task, then, is to assess what your data are saying in relation to your question; NVivo's contribution is to select and sort the data for you, often with a degree of complexity which would simply not be possible working manually. NVivo will not produce for you a neat p value (as if such should be considered adequate in any case)! As with any analysis (quantitative or qualitative), your results will only be as good as is allowed by the combination of your skill in asking the questions, your coding, and your capacity to interpret what is found.

Saving queries

The results of queries do not automatically update when you add more data or coding. Rather, queries in NVivo can be saved for re-running, either at a later time when you have more data available, or scoped to a different set of data or group of cases.

- When you are setting up a query, check the Add to project box at the top of the query dialogue. You will then be asked to give the query a name.
- Once you have set up the query, click OK to simply save the query. To run it and save it at the same time, click Run.
- Create folders in the Queries area (Navigation View) to store different kinds of queries. To make a new query which will be stored in a particular folder, ensure that folder is selected when you choose to create it.

Results of queries

Results of queries (i.e., the text that is found) can be previewed or saved as a node. If you preview first, you can then choose to save. If you simply give a name for the results and don't change the default options, the node will be saved in the Results area (under Queries in the Navigation View). Nodes saved in the results area cannot be modified, although you can code on from them to a free or tree node. Alternatively, if you want to 'clean up' the results by adding additional context, or by removing unwanted finds, then copy the results node across into the trees or matrices area (Nodes navigation view) where you will be able to modify it.

As with other 'storage' areas in NVivo, you are able to create folders within the Results area if you want to sort the results nodes you have been making. Generally speaking, however, the results nodes are intended for rapid perusal and then deletion; move the ones that are interesting enough for you to want to keep into the main coding system.

It is *not* a good idea to generate new results before you have reviewed the last set! Beware of building up a large number of results nodes.

Iterative searching

Because you can always save the results of your queries as a node, they can be reflected upon, and used iteratively in further queries. You might start with a simple question, perhaps about the level of association between two categories, but then go on to seek further clarification or detail. Is it true for everyone? Does it depend on some other factor also being present? What characterizes the cases where this is not true?

- Run the first search with a different scope, to see whether this association holds only for some subgroups within the sample.
- Enter the saved results from a first query as a data item in a further query.
- Save a matrix node as a tree node to run it against another set of factors (i.e., to generate a 3-way matrix result).

As you progress with your project and grow in sophistication in using the query tools, you will be able to make your questions increasingly targeted, detailed and specific – providing your node system and project structure allow it.

Keeping track of conclusions

Make a practice of immediately recording what you learn from each investigation you conduct, even if it 'doesn't produce anything' (nothing is always something). Making these records as you go will save you enormous amounts of time, in three ways:

- you won't find yourself having to run the same query again, because you can't remember what it found;
- recording what you learned from the query will prompt you to ask further questions, and so facilitate your searching and analysis process;
- the record provides the basis for your results chapters. It's always easier to start writing results when you have something already written down, regardless of how 'rough' it is.

Avoid printing off endless text reports from your data – that is just a way of deferring thinking, and the mounting volume will quickly prove daunting.

You might choose to record your emerging understanding in a special memo in NVivo, where you are able to use see also links to the results nodes or particular passages which hold the results or the models which display them. Alternatively, create a results document in Word, and use headings liberally to identify the bits and pieces of ideas you are recording. Drop your ideas (notes or quotes) under a heading, and when you think you're done on that topic, try turning them into prose. The headings will allow you to use both the document map and the outline view to quickly access or sort the topics in your growing report.

STARTING THE JOURNEY …

– with your questions

This is an excellent time to check back through your journal and other memos to identify

- the original questions you set out with, and
- questions which have arisen as you have been working through the texts.

In the context of your project goals, try to locate these questions within an integrated framework, to give direction to your investigations.

Spell out your questions in ordinary language (Richards, 2005: 152); write them down; identify the concepts (nodes) that are involved. Then think through, from your questions, what this implies about how those nodes might be associated.

This will lead you to the search and query tools that will help to answer them. NVivo attempts to write the queries you will use, also, in plain English.

– with your memos

If you have been memoing as you worked through your documents, your memos will now hold a rich storehouse of ideas to garner and exploit. They will help you sort out where you are going, what steps you can take next, and at the end of the process, they will help you write the story of your analysis and justify your conclusions.

- Review your case notes to prompt ideas about how you might approach a within-case or cross case analysis.
- If you have coded your memos as well, use a simple coding query to identify what thoughts you have recorded on a particular topic (look for a node, in memos only). This may well serve as a launching pad for a deep analysis of that concept.

– with a puzzle

'Kick-start' your analysis by "generating a puzzle by early inspection of some data" (Silverman, 2000: 135). Silverman provides an example, where a recipient of advice does not respond to it in any way. He looks for other comparable situations, and finds that this arises where the advice has not been sought by the recipient (typically an adolescent), where the recipient might perceive it as having a disciplinary intent, and where there are other people present. He then puzzled about what this silence might achieve … .

Silences often generate puzzles in qualitative analysis, as do contradictions. Why does Mr Brown never say anything about his son, nor Ms Green explicitly acknowledge her father's point of view (Poirier and Ayres, 1997)? These are not questions that can be answered with coding or matrix building, but rather the reader "must *listen to and hear the silence*" (Laub, 1991: 58, quoted by Poirier & Ayres) to discern a story which is often not heard even by the narrator.

Missing categories generate puzzles, similarly (Singh & Richards, 2003). When Supriya Singh talked to families about their changing relationship with banks when the banking system was deregulated in Australia, they talked instead about money in their marriages – but not about issues of gender, power or equality with regard to money in marriage. Singh moved for guidance from the banking literature to the literature dealing with the sociology of money, anthropological understanding of cultural rituals, and information science. She then re-investigated her data (in NUD*IST at the time). Her reanalysis led to the concept of 'ritual information', where information about the joint bank account was channelled so that discussion of power and equality was blocked, while aspects of jointness and sharing were emphasized (Singh, 1997). This emergent theory about the social shaping of money was then tested in further projects.

– with disciplinary insights

Singh's experience points to an important role that disciplinary literature might play in assisting you to make sense of your data. On the one hand, the literature in your discipline might provide useful sensitizing concepts or theoretical insights to pursue (Charmaz, 2006; Patton, 2002). On the other, the disciplinary literature might be analyzed as a source of data in itself (Caron & Bowers, 2000):

> Dimensional Analysis[1] [of the literature] ... becomes a mechanism to do a comparative analysis with what is developed from interviews. So for example, in my study of caregiving, it allowed me to say that the researchers and professionals conceptualize and study caregiving in this way, the concept of caregiving (in the literature) has the following dimensions, integrates the following assumptions. In contrast, interviews with caregivers suggested that the concept of caregiving was understood as a fundamentally different thing. So then you are in a position to say that the professionals may design interventions based on their understanding of caregiving, but since it is at odds with the experience (in the following way) the intervention is likely to miss the mark. (Barbara Bowers, personal communication, 2nd August, 2006)

What are the goals of analysis in the disciplines within which you work? What kinds of theoretical insights are valued? How might existing theory be used to inform your analysis? Your answers to these questions will influence the place literature assumes in your project: the point at which you focus your attention on it, the way you approach it when you do – whether you study it early on or whether you complete an analysis of your own data before working in depth with others' contributions; whether you simply peruse what others have written or whether you subject that writing to detailed analysis.

- To compare within NVivo, make a set of all the literature sources you have used, and another of your interviews (or other data sources). Sets can be used then to directly compare what your literature says with what your other sources say, about particular topics (nodes) – use a matrix coding query to do so (put nodes in the rows, sets in the columns and AND as the matrix).

Insights gained from archival, historical or other documentary materials can be used similarly, either to alert you to concepts and issues to attend to in the data (as in some discourse approaches) or to provide a basis for comparison.

GOING FURTHER WITH CASES

Some projects will focus around one case, but even where there are multiple cases, one of the goals of (and reasons for) taking a qualitative approach, typically, is to develop a deep understanding of each of those cases before bringing them together in either cross case or variable (theme)-oriented analyses. Advances in knowledge stem from context-dependent investigation, such as is found in detailed case analysis (Campbell, 1975; Flyvbjerg, 2004). "Virtuoso" expertise is then developed through intimate familiarity with (eventually) thousands of concrete cases within an area of practice or knowledge (Flyvbjerg, 2004). It is through practical experience with these multiple cases, over and above knowledge of general theories and rules, that one gains nuanced understanding of a topic and its related issues.

Cases identify the units of analysis in most research (*cf.* Chapter 3). The methods of within case and cross case analysis are foundational to a range of methodologies from phenomenology to evaluation research. Throughout this book, I have emphasized the importance of understanding and learning from each case as you worked though it to gain an understanding of 'what is going on here', whether or not you would describe your project as a case study. Now it is time to build on and test a consolidated understanding that either accounts for all cases, or clarifies (and perhaps explains) any pattern of differences. The tension is to balance uniqueness and generalization; the particular with the universal; to not lose localized understanding in the 'smoothing' of individual differences.

Within-case analysis

The purpose of within-case analysis, as noted above, is to develop a deep understanding of a particular case. Any query can be set up to run within a single case: choose to run the query In: Items in Selected Folders > Select > Cases and then select the case you want to investigate. This will limit the results to items coded at that case.

- Use a Simple Coding Query to review what a node looks like within a particular case, e.g. if you need to explore a deviant or negative case to see why it is different.
- Search for text within a case to determine whether or how a particular expression was used by someone in that case.
- Use a Matrix Coding Query to undertake within-case comparisons or to look for within-case associations of nodes.

In a study of a decision to outsource the meter reading activities of a publicly-listed Australian energy company, James Hunter compared what different people involved in the company (decision makers, decision implementers, employees and contractors) said about a range of issues (actions taken, feelings evoked and views developed) at various stages through the process: leading into the decision to outsource, immediately after that decision (less than 1 year) and into the medium term (2 to 4 years). This analysis was then repeated for a contrasting type of business, but each case (company) first needed to be understood on its own (Hunter & Cooksey, 2004).

Cross case analysis

If you have been writing a summary or building a model of each case as you completed your initial coding of it, then this is a really good time to read through just the set of those summaries or to review those models, in a sense treating these now as your data – noting recurring themes, odd discrepancies, significant concepts. Now you are indeed ready for cross case analysis!

The dual goals of multicase (or cross case) analysis are succinctly expressed by Miles and Huberman (1994: 172):

> One aim of studying multiple cases is to increase generalizability, reassuring yourself that the events and processes in one well-described setting are not wholly idiosyncratic. At a deeper level, the aim is to see processes and outcomes across many cases, to understand how they are qualified by local conditions, and thus to develop more sophisticated descriptions and more powerful explanations.

The first of these goals is met by a case-oriented analysis, where the complexity of individual cases is retained as they are compared in the search for both unique insights and common patterns. The second is met by variable-oriented analysis, when groups of cases are compared based on common attributes, or the presence of common coding. You may already have done some of this using attributes (Chapter 6); strategies for variable-oriented pattern-seeking of this kind will be elaborated further, below. Typically, both the fine-grained and the smoothed types of analyses are needed for theory development.

Queries to support cross case analysis always involve using a Matrix Coding Query. Whereas analysis based on groups of cases (comparing those with a different values of an attribute, or all the responses to 'x' situation compared to 'y' situation) gives an averaged result, in cross case analysis the comparative focus is on the individual cases, with their uniqueness preserved.

- Comparing cases is similar (in computing principles) to comparing attribute values. This time, however, set up your Matrix Coding Query with the required cases in the rows, and free or tree nodes or other items of interest defining the columns. As with the matrix query using attributes, you will search for content of rows AND of columns. This will find the text for each specified item (free node, tree node, or other) separately for each included case, and display it in table format, allowing you to compare across cases.
- You might refine your query by scoping to a particular folder or set of documents, for example, to compare cases on a range of factors at the beginning of the study or intervention (Time 1). In this case, you would set up your query with cases in the rows, nodes in the columns, and scope to the folder or set of Time 1 documents.
- If you want to refine your analysis by comparing what has been said or what happens at different time phases in the life of an organization, or through repeated interviews with the same participants, use time-based

codes or sets to identify the columns in a matrix (with cases in the rows). By scoping the query to a particular node (In: Selected Items > Select [Node]) you will have a comparison of how each case progressed over time for that particular issue/topic.

Thus, with cross case analysis using a matrix coding query, you are able to:

- Compare cases on a specific factor, and then refine to consider, say, over time/phase;
- Examine and determine the significance of patterns of association in codes, for example, seeing how many (and which) cases have one or other or both of two nodes present (or three or four), and then reviewing the text at those nodes on a case by case basis.[2]

In addition:

- From a case by node matrix, using the filter button (RMB ▼) on a column in the resulting matrix, see for whom there are 'hits' on particular items, and then view what each has to say about it;
- Generate a coding table for cases which can be exported as a case by variable matrix for use in a statistics program.

When data are sorted by case, in addition to seeing common patterns, you can readily identify instances where individuals go against a trend. These cases might be outliers on a statistical measure, deviant cases in qualitative terms, or cases where there is an apparent contradiction in the data from the different sources (Caracelli & Greene, 1993; Miles & Huberman, 1994). Whichever of these situations applies, they warrant focused attention, as it is often through exploration of such cases that new understanding is gained.

Profiling and clustering coding for cases – seeking phenomenological 'essences'

The term phenomenology has been applied broadly from philosophy to a variety of research methods frameworks (Patton, 2002). A common feature of phenomenological approaches, however, is that the researcher focuses their analysis intensively on individual cases before seeking common themes.

Husserl gave the name 'phenomenology' to research which aims to study reality (a pure phenomenon) as it is individually experienced (Groenwald, 2004). He argued that people could know with any certainty only that which they themselves had experienced, that objects in the external world do not exist independently of consciousness. All prevailing knowledge of a phenomenon has to be put aside (bracketed) so that it can be directly experienced in its basic, untainted form, in essential relationship to the conscious subject. "Phenomenology asks for the very nature of a phenomenon, for that which makes a some-'thing' what it is – and without which it could not be what it is" (van Manen, 1990: 10).

Culture (through language at least) will always intervene, however, so what is gained is never pure experience, but a reinterpretation of the phenomenon – a critical reflection which calls into question the current meanings attributed to the phenomenon. In its original form, then, phenomenology sought objective description of an object apart from the perception of the subject, but through first-person experience ('intentionality'). It is useful, then, for giving a beginning view of a phenomenon, and also as a way of critically refreshing one's view of a phenomenon as one proceeds in an inquiry process.

Currently phenomenology (particularly in North America) is most often seen as an attempt to discern the essence of an experience derived from the subjective accounts of those having the experience (e.g., Giorgi & Giorgi, 2003). The common understanding of, or meaning given to the experience that is derived from these accounts is therefore impacted by prevailing cultural understandings. This perspective contrasts with the earlier European traditions of Husserl and Heidegger: rather than being objective and critical, it is subjective and uncritical (Crotty, 1998; Schwandt, 2001).

In Chapter 5, I wrote about using sets or models to identify nodes which appear to cluster together, either because they all related to aspects of a single broader concept or theme, or because they were linked in a relational (theoretical) sense. The phenomenologist may choose to employ and extend this strategy of clustering concepts to represent 'themes' within a case (i.e., to identify 'essences' of the phenomenon).

- Chapter 4, Modeling a case, provided an example of how this might be achieved using the modeller.
- Alternatively, cluster nodes for a particular case using as many sets as are needed for that case (identify the sets with both the case and a cluster name).

Smith and Osborn (2003: 72) note that "this form of analysis is iterative and involves a close interaction between reader and text" so that the clustering is driven by the data rather than the researcher's assumptions.

This process may be repeated for each case, working toward a composite picture to capture all essential themes regarding the structure and meaning of the experience. As the range of core themes is reduced and a common picture is synthesized, each case is examined for fit and differences noted (Giorgi & Giorgi, 2003).

- Make sets (of nodes) to represent these common composite themes.
- Use these sets in a case by sets matrix, to illuminate commonalities and differences across the cases. Use the matrix filter option to quickly see which cases are exceptions for a particular aspect of experience (then explore why they are different).

Now record what you've done, and the thinking and understanding it has prompted!

GOING FURTHER WITH CONCEPTS

> ... the move from coding to interpretation involves playing with and exploring the codes and categories that were created. Dey (1993) provides many ideas about how you can go about doing this. He suggests that once data are displayed in a coded form, the categories can be retrieved, split into subcategories, spliced, and linked together. Essentially, the codes and categories you have selected should be used to make pathways through the data. (Coffey & Atkinson, 1996: 46)

By now you should be well practiced in applying strategies for reviewing concepts stored at nodes (*cf*. Chapter 7). The issue now is to use nodes to launch more analytical thinking, and to keep important nodes within analytic focus, using memos to hold those developments (Strauss, 1987). Whether the nodes considered should come together to define a single 'core' concept or basic social process, or whether there might be more than one without losing the central focus of the proposed theory is a matter of contention among grounded theorists (Charmaz, 2006). "Often it is not possible to give a 'big picture' unitary account without making it so general it is trivial. Showing *why* the data don't come together under one theme, and what explains divergences, may be far more powerful than describing the theme" (Richards, 2005: 132).

As you go further with concepts you will not only identify those which are more central to your analysis, but you will see more clearly their multifaceted character, their dimensions – much as a table has length, breadth and height, or obsession has dimensions of focus, passion and drivenness – or perhaps that these 'concepts' are themselves dimensions of a 'bigger' phenomenon. Seeing concepts as having dimensionality provides understanding that any phenomenon is complex – and leads into analysis:

> Dimensionality thus calls for an inquiry into its parts, attributes, interconnections, context, processes, and implications ... this defines analysis as the designation of these and their ultimate integration, providing an understanding or theory of "all" considerations seen as involved in the phenomenon and as constituting the "whole" of it. (Schatzman, 1991: 309)

This 'whole', however, is subject to change based on situated requirements. The process of analysis and definition is ongoing and developmental.

Working your nodes in NVivo

Choose a node which appears to be quite central to your current thinking about your topic. This might be selected because it is common to many documents, or because it frequently appeared as a focal point in models, or because you have written extensive memos around it.

- Review the node using the strategies suggested in Chapter 7 – reading through, writing a description, recording any further ideas and questions prompted by it in a memo, refining the coding.
- Provide the node with a job description (What is its purpose? How is it meant to achieve that?) and then subject it to a performance review (Bazeley & Richards, 2000).
- View it with coding stripes visible. Depending on the node, you might choose to view stripes for all nodes, or refine your view to selected coding stripes (e.g., for people referred to in this context, or a set of consequences) – or both. Viewing in this way may prompt you to run a coding query to check more precisely the relationship between this node and another (see below, under *Exploring associations*), or a matrix coding query to explore possible differences associated with, say, contextual factors or conditions or consequences.
- Investigate the issues raised by your earlier review (the ones you've noted in the memo). Write about what you are learning as you do this – doing so will prompt further questions, exploration and understanding to add to your conclusions.
- Create a model and place this node at its centre. Ask to see associated items (particularly any relationships). What other nodes (or other project items) should be linked to it? Again, write about what you are learning and use that writing to prompt further thinking. If a question arises as you are writing, explore or check it immediately.
- Repeat these processes for other nodes which appear to have a potentially central role in your project.

If you have a number of nodes which focus around a core idea (which might be referred to variously as a pattern code, metacode, axial code, selective code, etc.), try creating what Lyn Richards terms a 'dump node' with them (copy each of them and merge into a new node, or if they are in a concept-based set, create the set as a new node). You might add in the results of some text searches as well, just to make sure you have all possible aspects included. Now:

- Scan the contents of that larger node to gain a sense of its overall 'shape' and dimensions. Remove (uncode) any clearly irrelevant passages. Now scan it again, showing coding stripes for the original set of nodes. Do your original nodes adequately reflect the dimensions of this broader concept?
- Try using this new metacode in a series of matrix coding queries with attribute values and with other sets of nodes to see how it varies depending on who is the source of the comments, or the conditions under which they arise, or on what strategies are being applied, or what the consequences are. This may confirm your original subcategories, or it may reveal a new way of seeing (new dimensions in) this concept (Schatzman, 1991).

- Use of matrix coding queries will also clarify the relationships of this potential core category with other categories and concepts – assisting you to build a 'conditional matrix' to reflect how the conditions under which it is being expressed relate to variations in its expression (Strauss & Corbin, 1998). Understanding what brings about the different realizations of a phenomenon helps to create understanding of the phenomenon itself (Peräkylä, 2004).
- Again, repeat these processes for other sets or metacodes which also have a potentially central role in your project.

Which nodes (original single nodes, or metacodes) still 'stand up' (have the best explanatory power) after going through these examination processes? These are categories that will move you forward in your analysis as you look for ways they, with their dimensions, might link together in a model or theory.

Clustering for concept analysis

Concepts might be clustered in order to identify (depending on your perspective) broader dimensions across them, or essential elements within them. Do this by looking at the co-occurrence of nodes across all your cases, to see which were used in the same context.

- Set up a matrix of nodes in which all relevant nodes are listed in both the rows and the columns of the matrix (use original nodes rather than composite nodes, entering them *in the same order* in both rows and columns).
- Your choice of matrix (i.e., how the nodes will be combined) will depend on how you set up your documents and applied your coding.
  - If you used a table format and divided your text into units to which you applied coding as a whole, then choose AND. If you used free-form text and coded part or whole paragraphs, then choose NEAR-In The Same Paragraph. These will give you a pattern of 'tight' associations, showing which were used in the same context.
  - Alternatively, run the same query, but choosing NEAR-In The Same Scope Item (where your scope is to Documents) as the Matrix, to see where the nodes co-occur across whole documents (which, in a phenomenological study, would be equivalent to cases).

In this type of query, the numbers in the downward diagonal represent the totals for those nodes (these can be ignored for now). The numbers in the other cells tell you how often the nodes represented by that row and that column are used either together or near each other. For those taking a phenomenological approach, you might look at the frequencies of these combinations to identify groupings of nodes, and then compare how these match up with your more intuitively drawn clusters.

For the statistically minded, and depending on the nature of the data at hand, multi-dimensional scaling techniques (MDS) might be used to arrive at a mapping of codes which have been derived from qualitative coding, so as to generate what might variously be called major themes, dimensions or meta-codes (Bazeley, 2006; Ryan & Bernard, 2000, 2003). The frequency with which pairs of nodes are used in the same context (ascertained using an AND or NEAR matrix of nodes by nodes) can be transferred to a statistical program as a similarity matrix. MDS plots the position of nodes on a small number of dimensional axes, allowing the researcher to interpret what those underlying dimensions are. Two- or three-way dimensional plots also allow for observation and interpretation of clusters of codes.

In the larger study (involving 295 cases) from which the survey data in the Researchers project was drawn, descriptions of researchers performing in the eight different ways were coded for any of 56 different descriptors. A similarity matrix was developed from the frequency of association of codes in describing any one of the eight different 'brands' of research performers, and transferred to a statistical program able to read matrix data. MDS is being used with this data to identify key dimensions of research performance.

Typologies and taxonomies – classifying concepts

Classification systems and typologies fall partway between description and theory. While some are very enthusiastic about them, seeing them as a way of clarifying the territory of a subject or phenomenon, others see them as little more than description that doesn't offer anything new.

A good classification system or taxonomy, essentially, brings order to a mass of concepts – as indeed you should have already experienced in using NVivo's hierarchical tree system to classify (or catalogue) your nodes. In doing so, it helps you to see the structure of the phenomenon or process you are investigating.

Ian Coxon, an industrial design student at the University of Western Sydney, has a passion for 'new mobility vehicles' (sustainable, semi-covered two or three wheeled vehicles). In taking a hermeneutic phenomenological approach to the design project, he has found that NVivo's hierarchical coding system was the perfect vehicle for displaying the structure of his (and others') experience of using these vehicles. His Taxonomy of Experience (ToE, outlined below) demonstrates a range of human factors that designers of these vehicles might consider. It also provides a model for further design-related investigative research (for a range of design projects).

Ian Coxon's Taxonomy of Experience[3]

The Body ~ Somatic Experience
  Sensorial stimuli – taste, sound-hearing, smell, touch, sight; Comfort
The Heart ~ Affective Experience
  Negatively valenced affect (7–1) – sense of revulsion ... lack of emotion
  Positively valenced affect (1–7) – excitement of danger ... mystical moments, joy, elation
The Head – Cognitive Experience
  Conation – reflective thought (doing) – heightened awareness, physical tension
  Cognition – reflexive thought (thinking) – personal identity, detachment
Out There – Contextual Factors
  Environmental/Regulatory/Social/Existential/Corporate factors

Rather than providing a comprehensive classification system, a typology is often built on the end points of a continuum. Two continua might be cross-tabulated (in NVivo, by using matrix queries) to generate new "potential categories" in a "logical analysis" that may provide new insights to check against the data (Patton, 2002: 468). The associated warnings are to beware manipulating the data (Patton, 2002), and to watch for what might be lost from the data (Richards, 2005). The initial benefit, until patterns are modified and/or reconfirmed, is as a prompt to new ideas. Typologies can also be useful as a presentation device, especially when meaningful (indigenous) names can be attached to the various types.

GOING FURTHER WITH NARRATIVE AND DISCOURSE

Narrative has both content and form, and the two work together to communicate the meaning of experiences as understood by the teller. Whole interviews might comprise a mega-narrative, especially in life history research, or brief narratives might be found scattered throughout a record of interview. Narrative is built around a plot designed to convey the significance of the story of a person over time, in contrast to the cross-sectional perspective of discourse where the focus is more on language used than the person using it (Kirkman, 2002).

Briefer 'stories' might be better thought of as accounts rather than narratives. In these the teller is seeking to produce a plausible account of social events or social action, perhaps to explain their motives, or justify their action (Coffey & Atkinson, 1996). Accounts are culturally and socially situated, and so the analytic focus shifts to the vocabulary and language structures (e.g., idioms) used. By looking at how accounts are constructed, we can get a view of how social realities are constructed, along with the individual speaker's perspective (e.g., whether the scientist views results as 'luck' or the results of careful work).

Coding, marking off, annotating and linking text to identify structural features, literary forms and emotions, as well as content, all assist in identifying and interpreting the stories told and accounts given by research participants.

Narrative structure

Labov's model of structural analysis is most commonly referred to in discussions of narrative structure (Cortazzi, 1993; Riessman, 1993). This comprises six elements, although Cortazzi (1993: 47) notes that a coda may not always be present:

Abstract – summary of what the narrative is about
Orientation – time, place, situation, participants
Complicating action – sequence of events
Evaluation – significance, what does it mean to the narrator?
Resolution – what finally happened?
Coda – return to the present.

Alternative ways of defining narrative structures are described by Riessman (1993) and Elliott (2006). Elliott argues that Labov's structure is more useful for narratives encapsulated within a longer interview, rather than for whole longer narratives, and that the latter may be better described in terms of genre (tragedy, comedy, epic, fable, horror story, Gothic novel, melodrama), or in terms of the direction of the plot (progressive, regressive, or steady). Riessman found that for some of her transcripts, a method of hearing poetic stanzas (based on Gee, 1986; discussed also by Elliott, 2006) was the most effective way of identifying structure in those narratives. In their study of cancer patients, Mathieson and Barrie "... interpret the interviews as an unfolding of a central or prime narrative about a person with cancer and how this person has come to be in the place he or she is now. ... [The] narrators actively construct who they are by what they tell us in the interview, what they emphasize, and how they emphasize it" (1998: 583, 587).

Structure in a prime narrative

In analyzing the identity narratives of his urban Aboriginal participants, Reuben Bolt, a lecturer at Yooroang Garang School of Indigenous Health Studies and PhD student at the University of Sydney, identified the prime narrative for each, and then applied Labov's model of structural analysis to that prime narrative. Reuben illustrates using Mimi's narrative, which focuses on pride. Although the theme of racism is introduced in the complicating action, he notes that it is not the point of what was most important to her:

Component	*Line Number and Narrative*	
Orientation	*17:*	*What's important to me?*
Abstract	*18:*	*What is really important to me in my life is who I am,*
	19:	*what I believe in*
	20:	*and what my father told me.*
	21:	*In my life ...*
	22:	*(Could you tell me a little bit about that?)*
Orientation	*23:*	*When I was growing up,*
	24:	*as a young girl my dad always said to me "always be proud of who you are*
	25:	*because you are Aboriginal*
	26:	*and never, never, ever be ashamed of who you are*
	27:	*and always hold your head up high and walk tall*
	28:	*because our race of people is one of the most oldest races in the world"*
Evaluation	*29:*	*and we've always been very close as people and extended families.*
	30:	*and so I had a wonderful upbringing and*
	31:	*(I) never had a chip on my shoulder about who I am,*
	32:	*about being Aboriginal*
Complicating Action	*33:*	*... even when we went to school and other kids called us names and that*
Orientation	*34:*	*because they knew we were Aboriginal*
Resolution	*35:*	*but that never affected me because of what my father told me, what my father said to me.*
	36:	*I was always guided by what he had to say,*
Orientation	*37:*	*and he was a wonderful wise old man and ahh, I thank God that I had wonderful parents,*
Orientation	*38:*	*both mother and father (are) both Aboriginal*
Evaluation	*39:*	*and my brothers and sisters, we were all brought up to be proud of who we were.*

Given the typically discursive nature and length of narratives, reduction of the text in some form or another becomes necessary, for example, by extracting only those segments related to a critical theme (Riessman, 1993). In coding for structure, the user of NVivo may find it more appropriate to code on from text already coded for a critical theme, or to identify a complete brief narrative within a longer interview, rather than trying to identify structure in an initial pass over the full document:

- Start by creating a coding report for a document: Tools > Reports > Coding Summary – for a Selected Item (the source), with no extra information. This will provide guidance as to critical themes for that narrator in terms of number of passages for each theme, and the proportion of text coded at each.[4]

- Once a theme or themes have been selected for further consideration, gain live access to the text coded at the node just for that particular case using a simple coding query: New Query > Coding > Simple [selected node], scoped In: Selected Items [selected document or case], and save the results as a tree node. This will provide a sequenced view of all segments coded at a particular theme, with access to the context of each segment as needed and the capacity to add further coding or to uncode.
- Examine the text for structure and other elements of discourse, and then selectively code for these features. You might also use further coding queries, or selected coding stripes on the larger body of text, to examine the relationship between this theme and other components of the larger narrative.

Narrative as sense-making

Narrative conveys how the speaker makes sense of the events of which they tell. In Labov's structure, while the complicating action gives the chronology of events, "it is the evaluation that conveys to an audience how they are to understand the *meaning* of the events that constitute the narrative, and simultaneously indicates what type of response is required" (Elliott, 2006: 9). Elliott notes that narratives may involve distortions of facts, but that these distortions convey meanings given to events which provide valid understandings of lives in social contexts and they convey understanding of the cultural framework (intersubjective meanings) within which the individual operates.

The skills of literary analysis, in a framework of reader-response theory, were combined with insights from nursing and feminist theory by Poirier and Ayres (2002) in their analysis of narratives (both real and fictional) of family caregiving. Interview narratives were paired with literary narratives as they examined meaning in caregiving experiences in the light of alternative theories. In doing so, they demonstrated how "understanding proceeds through a constant movement between data and ideas" – ideas which come from disciplinary perspectives, previous research, theory, and general reading (Coffey & Atkinson, 1966: 153).

In NVivo:

- When reading the plot of a narrative, consider it in the context of both the details of that individual's life and the social and cultural context of that life. These aspects may have been noted via coding (especially where coding has been used to pick up the contextual and emotional elements of what is reported within any sentence or paragraph), in annotations or in detailed memos on the text, or using see-also links to other sources or memos. These linking strategies are especially relevant where what is being reported is coloured by previous events or future social or cultural considerations, or where the relevant contextual factor is not explicitly referred to by the speaker.

- Construct a (vertical) time line to represent the chronology of the narrative and the contextual influences on that chronology using a matrix coding query of time periods by events to provide the data. Note the significance of each event to the narrator, and to the listener. This interpretive activity could be done on paper, or in the modeler, using events nodes linked to a time line, with notes attached to those nodes to note their significance (and a memo for the model as a whole).
- Reflexively consider, also, your own response to the narrative, how your own history has impacted on that, and what that means for the way in which you are interpreting the narrative. Again, you are likely to do this primarily through use of memos.
- Use the modeler to visualize the pattern of events in their individual, social and cultural context.

Exploring metaphors, folk terms

"Metaphors are grounded in socially shared knowledge and conventional usage. Particular metaphors may help to identify cultural domains that are familiar ... they express specific values, collective identities, shared knowledge, and common vocabularies." (Coffey & Atkinson, 1996: 86). It is useful, therefore, to analyze the purpose of using a metaphor for the speaker, and the shared understanding it conveys.

Earlier (in Chapter 4) I suggested tagging such terms with a *metaphor* code as you work through your documents, as well as coding the content or meaning they convey in the context in which they are used. Now return to the metaphors you have identified, and review them in the light of your increased understanding of the culture, conventions and language of the people or situations you are studying. Ask what the speaker's purpose was in using the metaphor, what visual images does it portray – do these help you to see the world as they construct it? Do these terms speak of particular traditions or belong to a particular pattern of discourse?

- Explore intriguing or puzzling words, or metaphors using *text search,* and write about their uses. Scope text searches to examine use of words in different contexts or by different groups of people, or make a node from the text search results and put it into a matrix for comparison.
- Use the modeler to explore your metaphor. Identify the components of the image, and link those to data. Does it help to explain what you are seeing in the data? How far can you extend it before it loses veracity or meaning? Does it change shape, depending on the circumstances?

Shane's reference to "the sensuous feel of century-old newsprint" is suggestive of a romance, a passionate love affair with textured archival materials that began with his student project. The metaphor continues as he speaks of his love of handling personal papers and pictures, to unlock the historical secrets within.

Metaphors from the text or those you create may be useful for conveying aspects of your conclusions. They can convey patterns that connect findings to theory; they may have a richness and complexity that allows you to see new theoretical possibilities. Miles and Huberman (1994) suggest being playful (perhaps in groups) in experimenting with metaphor. For example, they developed the metaphor of *oasis* for the special-learning withdrawal room as a place of refreshment and refuge for children struggling with regular classroom learning environments. Metaphors, and also folk terms, conjure up rich and illuminating visual images in the mind of the hearer, but you need to exercise caution in not overextending an image.

Discourse – analyzing language in context

Discourse analytic methods (DA), of which there are multiple versions, generally derive from social semiotics – the analysis of signs and symbols. Analysis of signs involves studying the rules or forms of language, and the relationship between language and behaviour. In this latter sense, the analyst is often more concerned with the social problem (as reflected in language) than with the specific linguistic terms used. Like narrative, DA methods derive from a constructionist perspective; they are based in a symbolic interactionist framework.

Discursive psychology is "concerned with the role of talk and texts in social practices" (Hepburn & Potter, 2004: 185). Discourse is seen to be the primary medium of human action and interaction. "Rather than seeing such discursive constructions as expressions of the speaker's underlying cognitive states, they are examined in the context of their occurrence as situated and occasioned constructions whose precise nature makes sense to participants and analysts alike in terms of the social action these descriptions accomplish" (Potter & Edwards, 1992: 2).

The defining features of ***critical discourse analysis*** (CDA) "are its concern with power as a central condition in social life ... Not only the notion of struggles for power and control, but also the intertextuality and recontextualization of competing discourses in various public spaces and genres, are closely attended to" (Wodak, 2004: 199). The CDA analyst pays more attention than those working in discursive psychology to the social setting and historical context, incorporating fieldwork and ethnography as part of the method (Hepburn & Potter, 2004; Wodak, 2004).

Foucauldian discourse analysis (FDA) focuses on a problem, seeking to find the beginnings of related practices in historical archives, and asking how this came to be, rather than why (Kendall & Wickham, 2004). The past is used to understand the present, and as a springboard into the future. Historically created discourses (which may be more or less dominant in a society) – of science, medicine, and sexuality for example – serve prevailing institutional structures as subjectivity is shaped through the integration of power and knowledge. Reversal is seen to be possible through resistance which recognizes the possibility of alternative or counter-discourses (Seibold, 2006; Willig, 2003).

Researchers focusing on discourse employ a diverse but often interrelated range of methodological approaches. Suggestions for potential strategies which include use of NVivo to assist analysis include:

- Read the whole text (or relevant part of, say, a larger published text) and write a memo about the general sense of what it is doing. Then identify the object of interest, a particular aspect to explore in detail, and select relevant material from the text. Some use text search to identify potentially relevant passages for more detailed analysis.
- Locate these accounts in wider discourses, including theoretical perspectives for some, and (for FDA in particular) identify relevant archival material.
- As well as coding the major discourse(s), consider (code and memo) the construction of the object/account in terms of its action orientation (social purpose), terminology and other linguistic features, and subject positioning (where the speaker locates him or herself). (Grounded theory approaches to open and axial coding may be relevant here and below with regard to contexts and consequences.)
- Compare different versions of constructions (within and across cases).
- (FDA) Compare the historical construction with the construction of the participants in terms of major discursive and alternative constructions of the topic (matrix coding query, using nodes by sets of sources).
- Consider the context in which the different constructions are being used (matrix coding queries using nodes by attributes or nodes by nodes).
- Note (memo) the consequences in terms of limitations for action for the participant.
- Use models to show alternative constructions within their contexts, and showing their different consequences.

In the Researchers project one can trace discourses of performance, of duty, of romance, of play, but I note that no students refer to play – this is the preserve of experienced researchers only.

Conversation analysis

Conversation analysis (CA) typically focuses on brief examples of naturally occurring interaction to study patterns of talk as action and response (which is a further action), and so to identify the rules of conversation (Drew, 2003). CA emphasizes the structure of talk, rather than its purpose, and so is more concretely empirical (descriptive) than interpretive. It has been employed to analyze both ordinary and institutional (e.g., call centre, clinical) conversations (Peräkylä, 2004).

Drew (2003) describes how an inconsistency in what someone says (e.g., a strong denial followed by an exception) might at first be interpreted as attributable to some psychological variable in the speaker, but when a collection of extracts with this feature are considered, it begins to appear less as a psychological attribute and more like something generated by the interaction. Thus a pattern is revealed where the initial version of the retracted statement is always strongly or categorically stated, the recipient avoids endorsing the initial version (by silence or a minimal acknowledgement), and the subsequent retraction is given with the same lexical form but in explicit contrast to the initial version. These subsequent versions, then, are designed to be consistent with the initial version, but to allow an exception. Analysis of the setting in which these statements arise indicates that the initial overstated versions are necessary to achieve a social purpose which would not have been achieved with the weaker version if it had been given straight away. It can, however, be given safely once the moment requiring the stronger version has passed.

Drew (2003), Peräkylä (2004), Silverman (2000) and ten Have (1999) all provide guidelines for and examples of analytic strategies for conversation analysis. Specific ways in which NVivo may assist, beyond identification and coding of conversational units, are:

- Review a node which references segments of talk which are of interest (e.g., denials). Retrieve the context of the segment of talk to identify what led to it, the role the speaker takes through their talk, and the outcome of the talk (RMB > Open Referenced Source).
- Explore variations in sequences (adjacency pairs), such as a repair where misunderstanding has occurred, or response to an invitation, using NEAR or PRECEDING content – within # words, in an Advanced Coding Query, Matrix Coding Query or Compound Query.

USING NUMERICAL COUNTS WHEN ANALYZING TEXT

Content coding of text for quantitative analysis is a well established procedure, supported by increasingly sophisticated computer programming to parse complex sentences and generate maps of semantic networks, rather than simply count words (Carley, 1993). Numeric counts of instances of a theme within unstructured text are sometimes used as a proxy indicator of the importance of that theme for a qualitative analysis (Onwuegbuzie and Teddlie, 2003). One might count the number of times an issue is raised (number of references), or the number of people raising it, for example.[5] Many would argue, however, that one cannot assume a direct correspondence between simple frequency and importance. For example, Reuben Bolt of the University of Sydney (*cf.* above) noted that comments about racism were high in frequency in the stories told by his Aboriginal participants, but he identified other themes (such as pilgrimage, pride, the importance of family) as prime narratives in the talk of his Aboriginal participants. He argued that the narrative structure to these themes within their life stories, a structure that was missing from the theme of racism, demonstrated their higher relevance for those individuals.

Counts can be a useful supplement when considering some aspects of discourse. Bolt also noted that co-construction of narrative is important in the 'new ethnography' of Indigenous research where participants play an active role rather than being the subjects of research. He identified "researcher talking occasions" in his interviews, and then classified these as being general questions (e.g., to initiate talk), probing questions (to gain more discussion on an issue that the participant has introduced) or clarifying questions (designed to clarify that the researcher has accurately interpreted meaning). Those participants who dominated the talk in 'lecturer' style, to teach him rather than converse with him, more often prompted clarifications, while others prompted more probing questions. In an entirely different application, Anderson *et al.* (2001) used N4 to count lines between instances of argumentation strategies in children's discussions about various scenarios to test hypotheses about the snowballing effect of a child's introducing a particular strategy (the gap between instances diminished as it was adopted by others).

Wherever counts derived from textual data are used, care needs to be exercised in defining just what is to be counted (e.g., what counts as a 'researcher talking occasion'), and in determining whether a count of cases, references, or words is the most appropriate measure (Bazeley, 2004). The manner in which coding was applied to the text can also influence counts of passages and words (*cf.* Chapter 5).

GOING FURTHER INTO THEORY BUILDING

> We recognize that talking about theory or theory building can seem slightly daunting to some researchers. The thought that your data and the analysis of them has to use, contribute to, and make sense of or build theory can halt the research process altogether. We can think about theory in terms of having and using ideas, and this seems far less daunting. Everyone can use, develop, and generate ideas. (Coffey & Atkinson, 1996: 140)

Essentially what you are looking for in developing theory is to specify patterns and relationships between concepts. Theory allows for explanation and prediction. "The acts involved in [developing theoretical sensitivity] foster *seeing* possibilities, *establishing* connections, and *asking* questions. ... When you theorize, you reach down to fundamentals, up to abstractions, and probe into experience. The content of theorizing cuts to the core of studied life and poses new questions about it" (Charmaz, 2006: 135).

Schatzman (1991: 305) argued that theory building in qualitative research is an extension of "natural analysis" – the kind of thing we do in everyday life whenever we encounter a problem. Often theory built through qualitative methods is substantively based 'local theory' because it has limited generalizability. Even local theory can provide useful constructs and ideas to test elsewhere, and may contribute, through a series of studies, to the building of 'formal theory' of a more generic nature.

Exploring associations and investigating exceptions

You were first introduced to the idea of exploring associations between concepts as a way of reconstructing the links in fractured or sliced text. In reconstructing links, you ask for text that is coded by both of two (or perhaps more) nodes; for example, where you are looking at the connection between such things as a context and an action, or an action and who was involved, or an associated emotional response, where both are talked about at the same time. Often you will want these in series, in which case it is more efficient to use a matrix query (see below), but there are times when you will want to explore just one possibly interesting relationship, such as the role of stimulation in experiencing satisfaction with or enthusiasm for research, or whether institutional pressure creates insecurity – and how that might impact on future enthusiasm or performance. These kinds of queries are explored also using a Coding Query (Advanced), where you look for the text that is coded by *Node X*, AND coded by *Node Y* (*cf*. Chapter 5). This type of query will find all the instances where the text is coded by both nodes, allowing you to evaluate the nature of the association between them.

When you look for associations in this way, it is advisable in many instances to also check the meaning of a lack of association – what does stimulation look like when it doesn't prompt satisfaction or enthusiasm, or institutional pressure when it doesn't create insecurity. This is achieved by running the same kind of advanced coding query, but changing it to ask for text coded by *Node X* AND NOT coded by *Node Y*. This type of search will find all the passages coded by the first but not the second node, and so the question then is whether they look different from those in which both were present?

These associations are not always neatly connected. If I ask whether being stimulated by research is a necessary condition to finding satisfaction in doing research, a clear association can be found where the same passage is coded by both *stimulation* and *satisfaction* – but there are many more instances where participants talk about each of these separately. These instances need to be found, to determine whether there might be an association between them, even though each was spoken about at different times. And there are many other potential associations where the text will never be exactly co-located, such as establishing a link between a background characteristic or experience and the process of *identification* as a researcher. In these instances, I would use a NEAR search, looking for times where both nodes were present for the same scope item, where the search was scoped either to cases,[6] or to documents which were equivalent to single cases. This will tell me whether the person whose text is coded for *satisfaction* also has text coded with *stimulation*, or whether the case in which *xxx* was present also experienced *yyy*. It is important to always carefully evaluate the results from NEAR searches, as the simple presence of both codes in the same source or case does not necessarily mean they bear any theoretical relationship to each other. Can their association be interpreted as being meaningful in terms of the question being asked? Particularly when the co-occurrence may be anywhere across a document or case rather than, say, within a paragraph or speaking turn, there is an increased likelihood of chance associations being included in the query results.

Again, it is useful (if not essential) to explore the negative cases – those cases where the association is *not* found. With NEAR searches, this becomes a little more difficult than finding the simple difference between one node and another (as outlined above); what you need here is to see whether *satisfaction* exists at all for cases where *stimulation* doesn't exist anywhere in their record, or whether *yyy* can be found when *xxx* is not present at all for that case. At the time of writing, there are two ways of achieving this (using *satisfaction* and *stimulation* as an example):

- Use the Grouped Find option to identify the set of Items Coded At, with the Scope selected for the node of interest (*stimulation*), and with the Range selected for Cases. Highlight all cases found, and Create As Set (e.g., cases coded by stimulation). Now choose to run a Matrix Coding Query, with *satisfaction* as the only item in the rows; the new set (*cases coded by stimulation*) as the only item in the columns; and NOT as the matrix operator.[7] The result will be any finds of satisfaction existing in cases not coded by stimulation, allowing an assessment of whether one can have satisfaction without stimulation, and if so, whether it has a different quality from satisfaction where stimulation exists.
- Alternatively, run a cases by nodes matrix, with *satisfaction* and *stimulation* as the nodes selected for the columns. The resulting matrix can be filtered on the stimulation column, to see more easily whether there is an associated pattern in the satisfaction column (in terms of both numbers and text). This method is rather tedious, however, if there are a large number of cases involved.

Exploring patterns

Matrix coding queries using contextual variables, including attributes, along with more detailed methods of cross case analysis, each play a part in identifying patterns of association and of difference. Matrix queries also make more efficient the checking of a series of associations, such as those between an action and a number of possible actors, or a series of issues and a number of possible responses to those issues. These strategies therefore play an important role in theory-building and theory testing.

An explanatory matrix of conditions, actions/interactions and consequences is seen as the centerpiece for theory-building in a grounded theory analysis (Strauss & Corbin, 1998).

> An explanation, after all, tells a story about the relations among things or people and events. To tell a complex story, one must designate objects and events, state or imply some of their dimensions and properties – that is, their attributes – provide some context for these, indicate a condition or two for whatever action or interaction is selected to be central to the story, and point to, or imply, one or more consequences. To do all this, one needs at least one perspective to select items for the story, create their relative salience, and sequence them. Thus "from" perspective, "in" context, "under" conditions, specified actions, "with" consequences, frame the story in terms of an explanatory logic embedded in [a] matrix. (Schatzman, 1991: 308)

Matrix-style pattern analyses are used extensively also in case studies and evaluation research (Miles & Huberman, 1994; Patton, 2002; Yin, 2003) to explore changes over time, comparative outcomes, rival hypotheses, impacts for different subgroups – and for communicating data. Comparisons might be planned as part of the research design, or may be based on differences incidentally observed in the process of working with the data. NVivo matrices have particular value in that they provide both numeric summary information and also access to the underlying text, thus maintaining a connection with the evidentiary database (Yin, 2003). The numbers will tell you *how many* or *how often* something varied; the text will tell you *how* the something varied.

The most commonly used one is an AND matrix, where you will be looking for direct connections between pairs of nodes (or other project items) insofar as both items relate to the same segments of text. Thus, in an evaluation study, one might search for instances where text segments coded at particular strategies are also coded with *facilitator* or *barrier*. As noted in Chapter 6, the AND matrix is always used with attribute values, and would also be typically used when considering contextual variables which might have been coded on the text (such as location, or stage of researcher development, or project implementation).

- Create a matrix query in which a variety of things you are considering (typically all children in the same subtree) define the rows, and the variable across which you wish to make comparisons (e.g., group, context, time, evaluation) defines the columns. Save the query in case you want to refine it in some way. Choose AND as the type of matrix. Choose the documents or cases for which you want to run it. Select preview, or to save the results.

- When the matrix query is run, from the RMB options or from the grid toolbar:
  - Select whether you want to view counts of sources (documents), cases, or some other option.
  - Turn on matrix cell shading to highlight where differences might lie.
  - If the display is 'around the wrong way', transpose it (though note, it will not remain transposed if you save the table).
  - If you selected to preview only, you might now choose to save the results (or to modify the query).
- Double-click particular cells to read the data behind the numbers.
- Export the table (numeric counts) to Excel or as a text file (use the File menu).
- Record what you have learned from the query results (and further questions generated)!
- If you want to use the results in a further query, or to export or print the text in the cells, copy the results node to the Trees area and work from the now hierarchically arranged display of the matrix results.

While it is possible to use NEAR or NOT in a matrix query, to do so is advisable only in limited situations (e.g., for clustering nodes), as the result is likely to provide an overwhelming amount of information.

If you start running a matrix query, and then decide you want to opt out of it, you will find a Cancel option under Tools > Query > Cancel Query. Note that some queries can take a long time to run – don't cancel out simply because the computer appears to have stalled. (Choose to set up large matrix queries last thing before you take a break!)

Drawing causal inferences

Establishing causation or drawing causal inferences through research has been a long contested goal in both philosophical and practical terms. Classical randomized control experimentation has been the gold standard means of establishing causation in psychological, social and health sciences in recent times, but there are many who challenge its supremacy (sometimes humorously! *cf.* Smith & Pell, 2003).

Miles and Huberman suggest, in contrast, that qualitative analysis provides "a very powerful method for assessing causality" because:

> Qualitative analysis, with its close-up look, can identify *mechanisms*, going beyond sheer association. It is unrelentingly *local*, and deals well with the *complex* network

> of events and processes in a situation. It can sort out the *temporal* dimension, showing clearly what preceded what, either through direct observation or *retrospection*. It is well equipped to cycle back and for the between *variables* or *processes* – showing that "stories" are not capricious, but include underlying variables, and that variables are not disembodied, but have connections over time. (1994: 147; emphasis in original, to denote issues discussed as features of causality)

Those working in program evaluation and with case studies have developed a range of strategies for establishing causal pathways, including working from an *a priori* logic model which stipulates a chain of cause-effect patterns over time, based on program theory (Patton, 2002; Yin, 2003). Time-sequenced pattern-matching techniques can then be used to check for evidence which supports the logic of each stage in the model, supported by techniques (some quasi-experimental) to test rival explanations. As with analysis of patterns and associations more generally, these types of analysis draw heavily on the matrix query functions of NVivo, using time variables (usually nodes or sets) and/or a range of contextual and consequential nodes. When experimental designs aren't possible, one might draw on the forensic strategy of working through a process of elimination of competing possibilities, to find the most likely cause from those that match the profile of the case (i.e., theory) (Patton, 2002; Scriven, 1976).

Critically, the researcher has to keep in mind the rules of causal logic in making their interpretations. Miles and Huberman note that people tend automatically to think in causal terms when they are checking out relationships which have been observed during a study – their suggestion to counter this is to try turning the causal relationship around, to see if it seems truer that way (does commitment lead to involvement, or involvement to commitment?). It is also necessary to look for intervening variables which might explain the relationship (or lack of one), or an additional variable impacting on both those being considered. Where the path cannot be firmly established, all one can do is suggest that these factors appear to have some association, without specifying (or implying) causality. Ultimately, establishing causality is an issue of validity (Brewer & Hunter, 2006) – and that relies on (a) the strength of your evidentiary database, (b) your skill in handling and interpreting it, and (c) the records, or audit trail, you have kept to demonstrate how your thinking has developed and been challenged. Keep those (dated) memos going!

Investigating exceptions

The comparative process lies at the heart of developing new ideas about concepts and their relationships, and of testing those ideas. Making comparisons helps to reveal the dimensions of the concepts you are using; to find the conditions under which they take particular shapes, and to explore the implications of patterns of association between them. But, associations are not always universal, patterns are not always neat; there will be exceptions.

In talking about exploring associations and patterns, above, I emphasized the necessity to consider also when 'found' associations are not present, and to explore the implications of that for the conclusions you were drawing. What about 'one-off' exceptions, though – the outlier, the deviant case, the one that isn't covered by your nascent theory, but perhaps isn't enough to disprove it also? Is this going to be the 'exception that proves the rule' or indeed, an exception which shows the rule isn't a rule after all? What implications does this case have, if indeed it 'breaks the rule' for no ascertained reason? Does it mean you need to start over in your theory-building, or can your theory tolerate the exception? In quantitative work it is standard practice to ignore extreme outliers, because they unnecessarily skew the data, and little if any effort is ever made to accommodate them. In qualitative approaches, however, it is generally considered that your explanations or emergent theory should be able to address the variations present in all cases.

Use this divergent case to challenge the ideas you have developed. Identify where it differs, then explore what might be associated with that difference.

- The first thing to check might simply be that you really do have an exception here – or has it come about because of coding error, or through some other methodological issue?
- The next is to check the memos that have been written about this case, to see if they hold clues as to why it might stand out as different from others.
- Is there something unique about the circumstances, or about the way this person has been classified (check attributes)?
- Is there some variation in the way critical concepts are expressed in this case (check content of nodes in comparison to others)?
- Is there an additional factor operating in this case, which might create the difference?

The final measure, of course, is to amend your theory to take account of the lessons learned from this case!

Putting it together

As you have been exploring patterns and relationships (associative and causal) in your data, you have been 'emerging' theory at a substantive level. You have had an active role in this process – it hasn't happened on its own. From these explorations:

- Seek to understand *how people construct* meanings and actions, as preparation for understanding *why* people act as they do (Charmaz, 2006);
- Check for intervening, mediating, or extraneous factors relating to your focal concepts;

- Build a logical chain of evidence, using 'if-then' tactics – propose an if-then and then check to see if the evidence supports it, noting that relationships have to make sense, and explanations need to be complete. "Qualitative analyses can be evocative, illuminating, masterful – and wrong" Miles & Huberman, 1994: 262).
- Explanations or conclusions need to have conceptual or theoretical coherence. This is a good time to draw, once again, on disciplinary literature, and to contextualize your work within that literature – perhaps to refute it, and certainly to go beyond it (Coffey & Atkinson, 1996).
- Use models to assist in clarifying what you are seeing, and eventually in synthesizing it so others can see it too. Return to your original model where you mapped your preliminary ideas about what you might find in your project. See how far you have moved in your understanding of the issues you are dealing with! You might now revise that model – or perhaps you need to create a whole new picture of where you see your project now. Attempting to present your conclusions in the form of a model at once reveals where links are missing, and forces you to think about possible solutions, or pathways.

Theories may provide causal explanations, but do not essentially do so: there is more than one model of theory-building, and more than one kind of conclusion to a study. Projects might, for example, also take the form of proposing 'ideal types' – "patterns or typifications constructed by the analyst out of all the actual cases observed ... intended to capture the key features of a given phenomenon" in a form which might then be generalized, through comparative analysis, to a wider set of similar phenomena (Coffey & Atkinson, 1996: 143), or of interpreting narratives of personal experience (Denzin, 1997). Thus, generalization can occur from both explanatory and interpretive theory. Either way, theoretical propositions need to be both conceptually dense and well-grounded. While they are grounded in local data, they can transcend that, nevertheless, to generalize to wider domains. As you build your theory, the evidence from which you form your own conclusions is the evidence you will present to convince your readers.

MOVING ON – FURTHER RESOURCES

No single book can ever cover the entire range of possibilities with regard to analysis of data, whether with or without a computer. My hope is simply that the ideas presented will stimulate you to think about ways of exploiting your data (and NVivo) more fully as a way to pay respect to your participants, and advance knowledge. My further hope is that the guidance given has assisted you to understand the NVivo software sufficiently well that you can now explore its capabilities further for yourself. In doing so, think about whatever it is that you want to achieve, think about how the software works, be prepared to experiment (make sure you save first!), and approach it with the attitude: "There must be a way I can make this … program do what I want it to do!" – and indeed, you will often find that there is (although, of course, you must not expect one program to do it all).

If you are looking for further resources, here's some places you might try. Several of these sites carry links to further sites and resources.

www.qsrinternational.com The developer's site carries updates to the software, teaching resources, links to trainers and consultants in many countries around the world, newsletters with articles about using the software, answers to frequently asked questions (FAQ), and a link to the QSR forum.

http://forums.qsrinternational.com Hosted by QSR International, these forums provide a site where users ask questions, and other users will offer guidance or suggestions to help you. The site includes a searchable archive of previous questions and answers.

http://caqdas.soc.surrey.ac.uk The CAQDAS networking project provides practical support, training and information in the use of a range of software programs for use when analysing qualitative data. The information provided includes comparative evaluations, links to developers websites, and access to trial versions of various software packages.

www.researchsupport.com.au My website carries resources (notes and sample projects) specifically to accompany this book, articles on using NVivo for mixed methods analysis, and information about training and retreat facilities at Bowral, in Australia.

www.lynrichards.org Lyn Richards' website has a growing list of resources available, including tutorial material based on the Volunteering project, and a post workshop handbook.

www.kihi.com.au Leonie Daws hosts a forum for trainers, and provides on-line training and support for people wanting to learn to use (or develop their use of) NVivo.

NOTES

1 Dimensional analysis is a form of grounded theory coding (Schatzman, 1991).
2 A query of this type is an alternative to using NEAR when you want to look at the association of nodes, as it ensures that you don't accidentally confuse different speakers in multi-person documents. (The latter involves an issue of display of results which will be addressed by QSR in future revisions.)
3 Ian breaks some of the 'rules' for structuring coding systems in the ToE, at the level of specific nodes. This is because it is driven by a particular purpose where the essential focus of the ToE is its structure, rather than the content of the specific nodes. When it is applied for an alternative design purpose, the specific nodes will be different, but in any case, all essential components of an experience need to be listed in a readily accessible way that can be immediately converted to other tools without relying on NVivo's querying options.

4 If the narrator is embedded within a group document, run a Case by Nodes matrix coding query to achieve a similar result. This will also provide immediate access to the relevant text for selected nodes (save the matrix as a tree node and delete all but the wanted theme nodes).
5 The number of references shown in the Nodes List View can be inflated by coding in memos, while that in the documents view is inflated by results nodes. If either of these is an issue, a more reliable count is obtained either by running a simple coding query scoped to documents only, or for a number of nodes, by creating a matrix of nodes AND a document set.
6 Again, this may be a situation where, if you have multi-case documents, you would be better off to use a case by nodes matrix (*cf.* Note 2).
7 It may seem strange that you are using a matrix query where you have only one item for each of the rows and columns: this is because this is the only kind of query in which you can select a Set of items as an item for the query.

References

Anderson, R. C., Nguyen-Jahiel, K., McNurlen, B., Archodidou, A., Kim, S., Reznitskaya, A., Tillmans, M., & Gilbert, L. (2001). The snowball phenomenon: spread of ways of talking and ways of thinking across groups of children. *Cognition and Instruction, 19*(1), 1–46.

Australian Securities and Investments Commission (ASIC) (2002). *Hook, line & sinker: who takes the bait in cold calling scams?* Sydney: ASIC.

Bazeley, P. (1999). The *bricoleur* with a computer: piecing together qualitative and quantitative data. *Qualitative Health Research, 9*(2), 279–287.

Bazeley, P. (2003). Computerized data analysis for mixed methods research. In A. Tashakkori, & C. Teddlie (eds.), *Handbook of mixed methods in social and behavioral research* (pp. 385–422). Thousand Oaks, CA: Sage.

Bazeley, P. (2004). Issues in mixing qualitative and quantitative approaches to research. In R. Buber, J. Gadner, & L. Richards (eds), *Applying qualitative methods to marketing management research* (pp. 141–156). Basingstoke, UK: Palgrave Macmillan.

Bazeley, P. (2006). The contribution of computer software to integrating qualitative and quantitative data and analyses. *Research in the Schools, 13*(1), 63–73.

Bazeley, P., & Richards, L. (2000). *The NVivo qualitative project book*. London: Sage.

Bazeley, P., Kemp, L., Stevens, K., Asmar, C., Grbich, C., Marsh, H., & Bhathal, R. (1996). *Waiting in the wings: a study of early career academic researchers in Australia.* Canberra: Australian Government Publishing Service.

Boote, D. N., & Beile, P. (2005). Scholars before researchers: on the centrality of the dissertation literature review in research preparation. *Educational Researcher, 34*(6), 3–15.

Boyatzis, R. E. (1998). *Transforming qualitative information: thematic analysis and code development*. Thousand Oaks, CA: Sage.

Brewer, J., & Hunter, A. (2006). *Foundations of multimethod research: synthesizing styles.* Thousand Oaks, CA: Sage.

Bringer, J. D., Johnston, L. H., & Brackenridge, C. H. (2004). Maximizing transparency in a doctoral thesis: The complexities of writing about the use of QSR*NVIVO within a grounded theory study. *Qualitative Research, 4*(2), 247–265.

Bringer, J. D., Johnston, L. H., & Brackenridge, C. H. (2006). Using computer-assisted qualitative data analysis software to develop a grounded theory project. *Field Methods, 18*(3), 245–266.

Bryman, A. (2006). Integrating quantitative and qualitative research: how is it done? *Qualitative Research, 6*(1), 97–113.

Campbell, D. T. (1975). Degrees of freedom and the case study. *Comparative Political Studies, 8*(1), 178–91.

Caracelli, V., & Greene, J. (1993). Data analysis strategies for mixed-method evaluation designs. *Educational Evaluation and Policy Analysis, 15*(2), 195–207.

Caracelli, V. J. & Greene, J. C. (1997). Crafting mixed method evaluation designs. In J. C. Greene, & V. J. Caracelli (eds), *Advances in mixed-method evaluation: the challenges and benefits of integrating diverse paradigms* (pp. 19–32). San Francisco: Jossey-Bass.

Carley, K. (1993). Coding choices for textual analysis: A comparison of content analysis and map analysis. *Sociological Methodology, 23*, 75–126.

Caron, C. D., & Bowers, B. J. (2000). Methods and application of dimensional analysis: A contribution to concept and knowledge development in nursing. In B. L. Rodgers, & K. A. Knafl (eds), *Concept development in nursing: foundations, techniques and applications*. Philadelphia: W.B. Saunders.

Charmaz, K. (2006). *Constructing grounded theory*. London: Sage.
Coffey, A., & Atkinson, P. (1996). *Making sense of qualitative data*. Thousand Oaks, CA: Sage.
Cortazzi, M. (1993). *Narrative analysis*. London: Falmer Press.
Corti, L., & Thompson, P. (2004). Secondary analysis of archived data. In C. Seale, G. Gobo, J. F. Gubrium, & D. Silverman (eds), *Qualitative research practice* (pp. 327–343). London: Sage.
Crotty, M. (1998). *The foundations of social research*. Sydney: Allen & Unwin.
De Gioia, K. (2003). *Beyond cultural diversity: exploring micro and macro culture in the early childhood setting*. Unpublished doctoral dissertation, University of Western Sydney, Sydney.
Denzin, N. K. (1997). *Interpretive ethnography: ethnographic practices for the 21st Century*. Thousand Oaks, CA: Sage.
Dey, I. (1993). *Qualitative data analysis: a user-friendly guide for social scientists*. London: Routledge Kegan Paul.
Drew, P. (2003). Conversation analysis. In J. A. Smith (ed.), *Qualitative psychology* (pp. 132–158). London: Sage.
Elliott, J. (2006). *Using narrative in social research: qualitative and quantitative approaches*. London: Sage.
Erzberger, C., & Kelle, U. (2003). Making inferences in mixed methods: the rules of integration. In A. Tashakkori, & C. Teddlie (eds), *Handbook of mixed methods in social and behavioral research* (pp. 457–488). Thousand Oaks, CA: Sage.
Flyvbjerg, B. (2004). Five misunderstandings about case-study research. In C. Seale, G. Gobo, J. F. Gubrium, & D. Silverman (eds), *Qualitative research practice* (pp. 420–434). London: Sage.
Forum: Qualitative Social Research [On-Line Journal] (FQS). (2002). *3*(2), http://www.qualitative-research.net/fqs-eng.htm (accessed 4/07/2002).
Frost, P. J., & Stablein, R. E. (1992). *Doing exemplary research*. Newbury Park, CA: Sage.
Geertz, C. (1973). Thick description: towards an interpretive theory of culture. In *The interpretation of cultures: selected essays* (pp. 3–30). New York: Basic Books.
Gibbs, G., Friese, S., & Mangabeira, W. (2002). The use of new technology in qualitative research. Introduction to Issue 3(2) of FQS. *Forum: Qualitative Social Research [On-Line Journal]*, *3*(2), 35 paragraphs, http://www.qualitative-research.net/fqs-eng.htm (accessed 4/07/2002).
Gilbert, L. S. (2002). Going the distance: 'closeness' in qualitative data analysis software. *International Journal of Social Research Methodology, 5*(3), 215–228.
Giorgi, A., & Giorgi, B. (2003). Phenomenology. In J. A. Smith (ed.), *Qualitative psychology* (pp. 25–50). London: Sage.
Grant, B. M. (2005). Fighting for space in supervision: fantasies, fairytales, fictions and fallacies. *International Journal of Qualitative Studies in Education, 18*(3), 337–354.
Groenwald, T. (2004). A phenomenological research design illustrated. *International Journal of Qualitative Methods, 3*(1), Article 4.
Hart, C. (1999). *Doing a literature review: releasing the social science research imagination*. London: Sage.
Hepburn, A., & Potter, J. (2004). Discourse analytic practice. In C. Seale, G. Gobo, J. F. Gubrium, & D. Silverman (eds), *Qualitative research practice* (pp. 180–196). London: Sage.
Hunter, J. D., & Cooksey, R. W. (2004). The decision to outsource: a case study of the complex interplay between strategic wisdom and behavioural reality. *Journal of the Australian and New Zealand Academy of Management, 10*(2), 26–40.
Hycner, R. H. (1999). Some guidelines for the phenomenological analysis of interview data. In A. Bryman, & R. G. Burgess (eds), *Qualitative Research* (Vol. 3, pp. 143–164). London: Sage.
Kelle, U. (1997). Theory building in qualitative research and computer programs for the management of textual data. *Sociological Research Online, 2*(2). http://www.socresonline.org.uk/socresonline/2/2/1.html (accessed 06/08/05).
Kelle, U. (2004). Computer-assisted qualitative data analysis. In C. Seale, G. Gobo, J. F. Gubrium, & D. Silverman (eds), *Qualitative research practice* (pp. 473–489). London: Sage.
Kendall, G., & Wickham, G. (2004). The Foucaultian framework. In C. Seale, G. Gobo, J. F. Gubrium, & D. Silverman (eds), *Qualitative research practice* (pp. 141–150). London: Sage.
Kirkman, M. (2002). What's the plot? Applying narrative theory to research in psychology. *Australian Psychologist, 37*(1), 30–38.
Kvale, S. (1996). *InterViews: an introduction to qualitative interviewing*. Thousand Oaks, CA: Sage.
Lieblich, A., Tuval-Mashiach, R., & Zilber, T. (1998). *Narrative research: reading, analysis and interpretation*. London: Sage.

Marshall, H. (2002). What do we do when we code data? *Qualitative Research Journal*, *2*(1), 56–70.

Mathieson, C. M., & Barrie, C. M. (1998). Probing the prime narrative: illness, interviewing, and identity. *Qualitative Health Research*, *8*(4), 581–601.

Maxwell, J. A. (2005). *Qualitative research design*. Thousand Oaks, CA: Sage.

Meijer, P. C., Verloop, N., & Beijaard, D. (2002). Multi-method triangulation in a qualitative study on teachers' practical knowledge: an attempt to increase internal validity. *Quality and Quantity*, *36*, 145–167.

Miles, M. B., & Huberman, A. M. (1994). *Qualitative data analysis: an expanded sourcebook*. Thousand Oaks, CA: Sage.

Mishler, E. G. (1991). Representing discourse: the rhetoric of transcription. *Journal of Narrative and Life History*, *1*(4), 255–280.

Morse, J. M., & Mitcham, C. (1998). The experience of agonizing pain and signals of disembodiment. *Journal of Psychosomatic Research*, *44*(6), 667–680.

Moustakas, C. (1994). *Phenomenological research methods*. Thousand Oaks, CA: Sage.

Onwuegbuzie, A. J., & Teddlie, C. (2003). A framework for analyzing data in mixed methods research. In A. Tashakkori, & C. Teddlie (eds), *Handbook of mixed methods in social and behavioral research* (pp. 351–384). Thousand Oaks, CA: Sage.

Patton, M. Q. (1990). *Qualitative evaluation and research methods* (2nd ed.). Thousand Oaks, CA: Sage.

Patton, M. Q. (2002). *Qualitative evaluation and research methods* (3rd ed.). Thousand Oaks, CA: Sage.

Peräkylä, A. (2004). Conversation analysis. In C. Seale, G. Gobo, J. F. Gubrium, & D. Silverman (eds), *Qualitative research practice* (pp. 165–179). London: Sage.

Poirier, S., & Ayres, L. (1997). Endings, secrets, and silences: overreading in narrative inquiry. *Research in Nursing & Health*, *20*, 551–557.

Poirier, S., & Ayres, L. (2002). *Stories of family caregiving*. Indianapolis, IN: Center Nursing Publishing.

Potter, J. &. Edwards, E. D. (1992). *Discursive psychology*. London: Sage.

Reason, P. (1988). Introduction. In P. Reason (ed.), *Human inquiry in action: developments in new paradigm research*. Newbury Park, CA: Sage.

Richards, L. (1998). Closeness to data: the changing goals of qualitative data handling. *Qualitative Health Research*, *8*(3), 319–328.

Richards, L. (2002). Qualitative computing – a methods revolution? *International Journal of Social Research Methodology*, *5*(3), 263–276.

Richards, L. (2005). *Handling qualitative data*. London: Sage.

Richards, L., & Morse, J. (2007). *Readme first for a user's guide to qualitative methods* (2nd ed.). Thousand Oaks, CA: Sage.

Richards, T. (2002). An intellectual history of NUD*IST and NVivo. *International Journal of Social Research Methodology*, *5*(3), 199–214.

Richards, T. (2004). Not just a pretty node system: what node hierarchies are really all about. Presented at *Strategies in Qualitative Research using QSR software*, Durham, UK, September. (www.qual-strategies.org).

Riessman, C. K. (1993). *Narrative analysis*. Newbury Park, CA: Sage.

Ryan, G. W., & Bernard, H. R. (2000). Data management and analysis methods. In N. Denzin, & Y. Lincoln (eds), *Handbook of qualitative research* (2nd ed., pp. 769–802). Thousand Oaks: Sage.

Ryan, G. W., & Bernard, H. R. (2003). Techniques to identify themes. *Field Methods*, *15*(1), 85–109.

Schatzman, L. (1991). Dimensional analysis: notes on an alternative approach to the grounding of theory in qualitative research. In D. R. Maines (ed.), *Social organizations and social process: essays in honor of Anselm Strauss*. NY: Aldine De Gruyter.

Schatzman, L., & Strauss, A. (1973). *Field research*. Englewood Cliffs, NJ: Prentice Hall.

Schwandt, T. A. (1997). *Qualitative inquiry: a dictionary of terms*. Thousand Oaks, CA: Sage.

Schwandt, T. A. (2001). *Dictionary of qualitative inquiry* (2nd ed.). Thousand Oaks, CA: Sage.

Scriven, M. (1976). Maximising the power of causal investigation: the modus operandi method. In G. V. Glass (ed.), *Evaluation studies annual review* (Vol. 1, pp. 120–139). Beverly Hills: Sage.

Seale, C., Gobo, G., Gubrium, J. F., & Silverman, D. (2004). Introduction: inside qualitative research. In C. Seale, G. Gobo, J. F. Gubrium, & D. Silverman (eds), *Qualitative research practice* (pp. 1–11). London: Sage.

Seibold, C. (2006). Discourse analysis: integrating theoretical and methodological approaches. *Nurse Researcher* (forthcoming).

Silverman, D. (2000). *Doing qualitative research: a practical handbook*. London: Sage.

Singh, S. (1997). *Marriage money: the social shaping of money in marriage and banking*. St Leonards, NSW: Allen & Unwin.

Singh, S., & Richards, L. (2003). Missing data: finding 'central' themes in qualitative research. *Qualitative Research Journal, 3*(1), 5–17.

Smith, G. C. S., & Pell, J. P. (2003). Parachute use to prevent death and major trauma related to gravitational challenge: systematic review of randomised controlled trials. *British Medical Journal, 327*, 1459–61.

Smith, J. A., & Osborn, M. (2003). Interpretive phenomenological analysis. In J. A. Smith (ed.), *Qualitative psychology* (pp. 51–80). London: Sage.

Stake, R. E. (2000). Case studies. In N. K. Denzin, & Y. S. Lincoln (eds), *Handbook of qualitative research* (2nd ed., pp. 435–454). Thousand Oaks, CA: Sage.

Strauss, A., & Corbin, J. (1998). *Basics of qualitative research* (2nd ed.). Thousand Oaks, CA: Sage.

Strauss, A. L. (1987). *Qualitative analysis for social scientists*. Cambridge: Cambridge University Press.

ten Have, P. (1999). *Doing conversation analysis: a practical guide*. London: Sage.

Tesch, R. (1990). *Qualitative research: analysis types and software tools*. London: Falmer.

van Manen, M. (1990). *Researching lived experience: human science for an action sensitive pedagogy*. NY: State University of New York.

Willig, C. (2003). Discourse analysis. In J. A. Smith (ed), *Qualitative psychology* (pp. 159–183). London: Sage.

Wodak, R. (2004). Critical discourse analysis. In C. Seale, G. Gobo, J. F. Gubrium, & D. Silverman (eds), *Qualitative research practice* (pp. 197–213). London: Sage.

Yin, R. K. (2003). *Case study research: design and methods* (3rd ed.). Thousand Oaks, CA: Sage.

Index